To Owen,
with warm regards,
Larry

EIGHTEENTH-CENTURY CHEMISTRY AS AN INVESTIGATIVE ENTERPRISE

Frederic Lawrence Holmes

Five Lectures delivered at the
International Summer School in History of Science
Bologna, August 1988

Office for History of Science and Technology
University of California at Berkeley
1989

Berkeley Papers in History of Science, 12
Uppsala Studies in History of Science, 6
Bologna Studies in History of Science, 1

Order from
Office for History of Science and Technology,
University of California
543 Stephens Hall
Berkeley, CA 94720
Price $16.00

ISBN: 0-918102-2
Library of Congress catalog card number: 89-51465

Content

PREFACE

The lectures that make up this book were delivered at the first session of the International Summer School in History of Science, which will be held every other year in Bologna, Uppsala, and Berkeley in rotation. The School brings together advanced students and professors to hear new interpretations and to consider new material about subjects with which they are already familiar. The School provides for two lectures in the mornings and for time in the afternoons to use collections in local libraries, archives, and museums. Each principal lecturer has available an hour a day for a week, which may be divided into formal lecture and informal discussion.

The School's organizers believe that the work of its lecturers deserves a wider audience than the School itself can provide. Hence they have established the series of which Professor Holmes' book is the first. The organizers esteem this bellwether not only as an important contribution to the historiography of early-modern chemistry, but also as an indicator of the high standards of research and presentation, the scholarly and friendly exchanges, and the quiet revisionism for which they trust the School will be known.

Holmes' sustained argument is that historians of eighteenth-century chemistry have missed an important part of their story by restricting their attention to the supposed antecedents and the detailed course of Lavoisier's chemical revolution. Holmes argues that early-modern chemists made very considerable progress in the identification and classification of substances and reactions without perplexing themselves over theories of combustion or attraction, or worrying where the phlogiston went. His prime example is the chemistry of salts formed from mineral acids and various bases. He shows that Lavoisier took over this body of knowledge virtually intact; it had its own principles and logic, and did not require dephlogistification to be useful. In order to show the development and power of the chemistry of mixed salts without assimilating it to Lavoisier's system, Holmes avoids the modern terminology that descends from the chemical revolution. With a little attention, however, the reader will master the operational meaning of regulus of antimony and vitriolated tartar, and have the profit and pleasure of learning a new language.

The lectures offer considerably more than the exposition of dead chemistry. They also review the historiography of the subject; offer a heroic example of abstention from whiggism; indicate new twists in the ties between science and technology; expose the pitfalls of what may be called "alchemical history;" and show in a new way the implacable force of numbers. These last two contributions require explanation. The alchemical historian believes in, and seeks, the philosopher's stone, the single key, the leitmotiv, that unlocks or runs through the life work of an innovative scientist or the course of a scientific revolution. The alchemical historian sublimes directly to the explicandum without passing through the state of searching for what needs to be explained.* As for the force of numbers, the multiplication of reagents and reactions pressed for a practical means of inventorying and accounting. In this respect the expert on mixed salts shared the problems of the lexicographer, encyclopedist, and classifier. These problems, originating in the continuing discovery of new plants, animals, lands, and societies, required solutions typified by Diderot and d'Alembert's *Encyclopédie*, Linnaeus' *Systema naturae*, and Macquer's *Dictionnaire de chymie*. Here the apparently parochial subject of mixed salts met the general themes of the Enlightenment.**

J.L. Heilbron

* Holmes gives a good lesson on the subject on p. 118. For other examples of alchemical history see J.L. Heilbron, "The electrical field before Faraday," in G.N. Cantor and M.J.S. Hodge, eds., *Conceptions of ether* (Cambridge, 1981), 187–213, on 189.

** Cf. infra, pp. 56, 124. This point is dilated in Tore Frängsmyr, J.L. Heilbron, and Robin E. Rider, eds., *The quantifying spirit of the eighteenth century* (University of California Press, forthcoming).

INTRODUCTION

The first International Summer School in History of Science, held at the University of Bologna between August 29 and September 9, 1988, was devoted to the topic "New perspectives in Enlightenment science." The present volume contains the texts of a series of five lectures on eighteenth century chemistry that I delivered during the first week of that School.

These lectures are heterogenous in character. The first is mainly a critical review of some of the recent historical studies in eighteenth century chemistry. The second and third are case studies of two important areas of eighteenth century chemical investigation, and are developed from an unpublished study that I wrote some time ago. The fourth applies a perspective based on the preceding cases to studies carried out by other historians on the origins of chemical industry. The final lecture is a commentary on certain aspects of the much–discussed chemical revolution. Illustrations of chemical laboratories and apparatus follow chapters 1, 2, and the Epilogue. Modern equivalents of the older chemical terminology used in the lectures are given in the Appendix.

Two central themes run through these lectures. One of them, indicated by the general title of the series, is that eighteenth century chemists prior to the events leading to the chemical revolution carried on a thriving investigative activity to which historians have given relatively little attention. The other theme, surfacing occasionally in all the lectures, but coming into full view only in the last one, is that the meaning of the chemical revolution must be reconsidered in the light of this earlier investigative activity. Taken together, the lectures provide a sketch of eighteenth century chemistry highlighting certain of its features that have not been displayed prominently in the existing literature. In time I hope to be able to develop this picture in fuller detail, and to integrate more familiar aspects of the subject with those upon which I have concentrated here. In the meantime I offer this preliminary treatment as an intentionally provocative stimulus to historians of science presently working in eighteenth century chemistry. If the positions for which I have argued bear up under further scrutiny, they will invite a substantial modification of the historical lines of inquiry currently prevailing in this field.

In several places in the lectures I have treated statements by senior historians, particularly Henry Guerlac and Charles Gillispie, as points of departure for the elaboration of perspectives diverging from their own. These are not intended as general criticisms of scholars whose work I admire, and who have served as mentors to my generation. Their positions were based on probing critiques of earlier views. If the further passage of time now allows us to test new perspectives against their interpretations, that could only happen through the assimilation of what they have taught us. Particularly to Henry Guerlac I owe my own introduction to eighteenth century chemistry.

The format of the International Summer School in Bologna permitted extended discussions among students and scholars steeped in many aspects of eighteenth century science. During the discussions that followed these lectures, questions were posed that complemented, supplemented, and in some cases ably challenged the viewpoints that I presented. In a brief epilogue I have responded to a few of the issues raised. Otherwise I have made only minor grammatical changes in the lectures as delivered.

I am grateful to the organizers of the International Summer School of the History of Science, Giuliano Pancaldi, Tore Frängsmyr, and John Heilbron, for the opportunity to take part in the inauguration of an institution that I believe can play an important role in the future development of the discipline of the history of science. The care with which Giuliano Pancaldi organized the first session of this school in Bologna, matched only by the graceful modesty with which he veiled the great amount of work that it entailed, made the events of those two weeks memorable for all who came together there.

1 VIEWS OF EIGHTEENTH CENTURY CHEMISTRY

Historians of science have found it difficult to view eighteenth century chemistry as anything other than the stage on which the drama of the chemical revolution was performed. So strong has the disposition been to identify the advent of the modern science with the chemical system established by Lavoisier between 1772 and 1789 that all earlier activity has been treated most often as a prologue to those climactic events.

A small set of achievements by his "predecessors," set against the background of a prevailing phlogiston theory treated variously as a qualified advance or an obstacle to progress, has come to comprise a canonical story culminating in Lavoisier's appearance on the stage. These earlier developments include especially the pneumatic experiments of Stephen Hales, which first enabled chemists to handle "airs" and to incorporate them into chemical changes; and the pioneering experiments of Joseph Black, which not only identified fixed air as the first species distinct from common air, but served also as the prototype for the quantitative methods afterward generalized by Lavoisier. Accounts of the discovery of further airs by Henry Cavendish and Joseph Priestley lead quickly to 1772 and Lavoisier's first experiments on calcination and combustion.

Without attempting to trace the origins of this historical account, I shall simply point out that its elements had solidified into a standard picture, presented with minor variations, in those general textbooks that dominated the emergence of the history of science itself as a discipline during the 1950s and 1960s. Such influential works as Herbert Butterfield's *The origins of modern science*, A.R. Hall's *The scientific revolution*, and J.R. Partington's *A short history of chemistry* differed in the details chosen, and their attitudes toward the phlogiston theory, but all told basically the same story. Since then there have been some changes of emphasis that have added further elements to the story. In particular, the recognition that Lavoisier had been occupied from the start with the nature of heat and of the gaseous state has led historians to give more attention to the prior history of these problems. The most recent general textbook chapter on eighteenth century chemistry, in Thomas Hankins' excellent *Science and the Enlightenment,* places as much emphasis on Joseph Black's discovery of latent heat as on Black's discovery of fixed air. Nevertheless, Hankins retains the basic pattern laid down by earlier textbooks. "The chemical revolution," Hankins observes, was "more the creation of a new science than a change in an existing one." It is the formation of this new science that dominates his succinct and lucid treatment of chemistry in the Enlightenment; and chemical activities that are not implicated in that process are barely mentioned.[1]

1. INTERPRETATIONS OF EIGHTEENTH CENTURY CHEMICAL THOUGHT

Historical discussions of eighteenth century chemistry that depart from this format have focused on two themes: atomic theories of matter, and chemical affinities. In an essay on "The development of chemistry in the eighteenth century," one of the few recent attempts to survey the broad intellectual outlines of that subject, Maurice Crosland in 1963 associated both of these themes with the pervasive influence of Newton on eighteenth century science. In the queries at the end of his *Optics,* Newton described atoms as "solid, massive, hard, impenetrable, moveable particles," and he explained a large number of chemical changes, with which he was familiar through his chemical and alchemical studies, by means of special attractive forces exerted between these particles at very short distances. Ideas about the ultimate constitution of matter more or less reflecting Newton's views continued to be prominent topics of discussion through the century, although they played little role in

1. Herbert Butterfield, *The origins of modern science: 1300–1800,* 2nd ed. (London, 1957), 191–209; A.R. Hall, *The scientific revolution: 1500–1800* (Boston, 1954), 325–338; J.R. Partington, *A short history of chemistry,* 3rd ed. (London, 1957), 85–151; Thomas L. Hankins, *Science and the Enlightenment* (Cambridge, 1985), 81–112.

chemical practice. The concept of chemical affinities, on the other hand, transmitted through the tables published in 1718 by Etienne Geoffroy, both enabled chemists to organize their knowledge of specific chemical reactions and guided further experimentation.[2]

Arnold Thackray, who elaborated on these themes in 1970 in his book *Atoms and powers,* claimed that a study of the influence of Newtonian ideas on eighteenth century chemistry was a corrective to the standard history of the subject defined from nineteenth century perspectives. The selection of matter theory and chemical affinities for special historical attention, however, is also influenced by the projection backward of a later state of chemistry. If Lavoisier represents the denouement of the canonical story, John Dalton's atomic theory justifies the focus on eighteenth century speculations about the atomic nature of matter. Thackray tacitly acknowledged as much:[3]

> It was within a peculiarly British context that Newton elaborated, and a century later Dalton transformed, those concepts of matter and its properties that were so strongly to influence research and debate on chemical units and mechanisms right down to the close of the nineteenth century.

The anticipatory nature of the historical attention given to chemical affinities is more subtle than the cases we have so far examined. At first glance, the fact that chemical affinity was a characteristically eighteenth century concept, that efforts to quantify affinities failed, and that Claude Louis Berthollet's concept of mass action undermined the eighteenth century idea of affinity at the beginning of the next century, seem to suggest that to fix on these ideas is a means of freeing ourselves from retrospective bias. Affinities are nevertheless, I believe, treated historically as somehow prophetic of the emergence of a future chemistry because of the underlying belief of earlier historians of science that a modern science should be based

2. Maurice Crosland, "The development of chemistry in the eighteenth century," *Studies on Voltaire and the eighteenth century,* 24 (1963), 369– 441, esp. 375–390. For Newton's chemical and alchemical activities, see Richard S. Westfall, *Never at rest: A biography of Isaac Newton* (Cambridge, 1980), 281–309, 357–376, and B.J.T. Dobbs, *The foundations of Newton's alchemy: or, "The hunting of the greene lyon"* (Cambridge, 1975).

3. Arnold Thackray, *Atoms and powers: An essay on Newtonian matter-theory and the development of chemistry* (Cambridge, Mass., 1970), 3.

on a coherent conceptual structure.[4] Historians have invoked chemical affinities as the nearest approximation to such a structure available before the chemical revolution. Thus Crosland wrote:[5]

> Tables of affinity [as distinguished from Newtonian ideas of attractive forces as the *cause* of affinities] are important in the history of chemistry in the eighteenth century for a number of reasons. In the first place, they reduced a multiplicity of chemical reactions to a series of simple sequences. Affinity tables did not of themselves constitute a system based on a theory such as that of attraction between atoms. They were based entirely on experimental evidence and constituted an important classification.
>
> When the phlogiston theory is viewed sympathetically in the light that it was a means of reducing a large number of facts to a coherent system, the value of the affinity tables is generally overlooked. Affinity tables, like a scientific theory, fulfilled the dual role of summarizing chemical reactions and predicting others.

Thus, Crosland implies, the eighteenth century had to make do with two surrogates for a true theoretical foundation: one theory that was wrong and another that was merely an empirical classification, but both of which substituted reasonably effectively for the organizing roles that at a later stage in development would be filled by the powerful theories emerging from the chemical revolution.

Concentration on those strands within eighteenth century chemistry that can be seen either as first stages in the movement toward the chemical revolution or harbingers of the theoretical framework of nineteenth century chemistry has created an essentially negative picture of eighteenth century chemistry as a whole. The subject is identified more by what was missing than by what was present. When Hankins writes, "Before 1750, chemistry could not be regarded as an independent discipline,"[6] he echoes a well–embedded historiographical tradition. A generation earlier Hall wrote that "chemistry was always an eminently practical science, as well as a branch of natural philosophy, and that it lacked (before Lavoisier) a coherent conceptual scheme."[7] Aaron Ihde began his

4. See, for example, Hall (ref. 1), 305.
5. Crosland (ref. 2), 383–384.
6. Hankins (ref. 1), 81.
7. Hall (ref. 1), 305.

book, *The development of modern chemistry*, in 1964 with the observation that, "Although chemistry as a modern science may be considered to have originated in the work of the pneumatic chemists following the discoveries concerning gases after 1750, it is necessary to understand earlier ideas about matter if the problems of the chemical revolution are to be properly comprehended."[8] The most unkind negative portrait of all was promulgated, ironically, by that self-proclaimed arch-foe of Whiggish history, Herbert Butterfield:[9]

> It would appear to us today that chemistry could not be established on a proper footing until a satisfactory starting-point could be discovered for the understanding of air and water; and for this to be achieved it would seem to have been necessary to have a more adequate idea both about the existence of "gases" and about the process of combustion. The whole development depended on the recognition and the weighing of gases; but at the opening of the eighteenth century there was no realization of the distinctions between gases, no instrument for collecting a gas, and no sufficient consciousness of the fact that measurements of weight might play the decisive part amongst the data of chemistry.

We may be tempted to dismiss statements such as those of Hall, Ihde, or Butterfield as relics of what we like to call the older positivistic historiography of science.[10] Hankins' view reminds us, however, that although we may have discarded the language of the earlier histories of science, we have yet to counteract effectively the images of many eras in past science that they have bequeathed to us. As I shall argue further on, recent historians of chemistry have probed more deeply and assessed more critically the crucial events within the canonical story; but they have done little to modify the overall contours of eighteenth century chemistry the story conveys.

Not all historians of chemistry have deprived early eighteenth century chemistry of the attributes of a modern science. One of the "older" historians, Eric Holmyard, could write in 1925 of the phlogiston theory that it "did more, in a single century, to make chemistry a rational science, than any of its predecessors had been

8. Aaron J. Ihde, *The development of modern chemistry* (New York, 1964), 1.

9. Butterfield (ref. 1), 193.

10. A germinal discussion for the perception that a new historiography of science has replaced that which was responsible for statements such as those quoted is Thomas S. Kuhn, *The structure of scientific revolutions* (Chicago, 1962), 1–2.

able to accomplish in a millenium."[11] It is undoubtedly because Holmyard had concentrated his extensive scholarly effort on chemical thought from antiquity until the eighteenth century that he was able to appreciate phlogiston from the perspective of what had preceded it, rather than to denigrate it in the light of what followed it. More recently, Robert Multhauf ended his book on *The origins of chemistry* at the 1730s with the assertion that, "By the early eighteenth century a *science* of chemistry had emerged from a matrix of natural philosophy, medicine, alchemy, and technology."[12] Although not free of presentist judgments, Multhauf was prepared by his long immersion in more ancient periods to see early eighteenth century chemistry as both a culmination and a foundation for the "reform" that came in later decades of that century.

The most conspicuous exception to the historiographic patterns that have dominated the treatment of eighteenth century chemistry is Hélène Metzger's *Chemical doctrines in France from the beginning of the 17th to the end of the 18th century.* Written in the 1920s and 1930s, these classic studies have influenced several generations of historians of science. Breaking deliberately with a history of chemistry that she perceived as devoted to the discovery of previously unknown chemical reactions, the confirmation of facts, brusque revolutions, and the triumphant successes leading to the present proud state of the field, Metzger set out to trace the "gradual evolution" of chemical theory characteristic of an earlier age. Her goal was "to establish the order and succession of different chemical doctrines."[13] She was not able to complete her entire plan, but in the two volumes she finished—the first treating mainly seventeenth century chemists, the second on Newton, Stahl, and Boerhaave—she penetrated the structure of the thought of these figures in ways that are still illuminating half a century later.[14] Although she did not escape entirely from the bias that these successive doctrines mark a long slow pathway toward the chemistry of Lavoisier, that orientation is muted and subordinated to her probing effort to understand

11. E.J. Holmyard, *Chemistry to the time of Dalton* (London, 1925), 56.

12. R.P. Multhauf, *The origins of chemistry* (New York, 1966), 349.

13. Hélène Metzger, *Les doctrines chimiques en France du début du XVIIe à la fin du XVIIIe siècle* (Paris, 1969), 7–13.

14. Metzger (ref. 13) and *Newton, Stahl, Boerhaave et la doctrine chimique* (Paris, 1930).

earlier chemical thought within its own intellectual framework.

Metzger was aware of the importance of the interconnections between "the evolution of doctrines, experimental discoveries, and technical invention" in chemistry. In her view, however, it was essential to establish the history of doctrines before attempting to explore these connections in detail. Consequently, although experimental practice is often mentioned, it remains always in the background of discussions of ideas. It was "above all the currents of thought" that she wished to investigate.[15] Metzger's approach to the history of chemistry complemented in some ways Alexander Koyré's approach to the history of mechanics and astronomy, and helped to shape the attitude prominent after World War II that the history of science is a part of the history of ideas. Now that that view has been rejected by many historians of science, attempts are being made to extract from Metzger's endeavor themes more attuned to current trends. Jan Golinski, for example, has recently emphasized that Metzger dealt with "orientations of mentalities;" and he has suggested that a fruitful alternative to her explanation of "formal conceptual structures" would be "to concentrate instead on the concrete practices involved in writing texts and producing discourses."[16]

The enduring impact of Metzger's work despite the major reorientations that have occurred in the history of science since she wrote is in part a testimony to the special insight she brought to bear on her subject; but it reflects also the modesty of the progress that historians have made since then in the study of eighteenth century chemistry. The subject has, during the last two decades, attracted less scholarly effort than one might expect from the central position long accorded the chemical revolution among the momentous events in the history of Western science.[17] Even Lavoisier

15. Metzger (ref. 13), 10, 12.

16. Jan Golinski, "Hélène Metzger and the interpretation of seventeenth century chemistry," *History of science, 25* (1987), 84–97, on 86, 92, 95. See also J.R.R. Christie, "Narrative and rhetoric in Hélène Metzger's historiography of eighteenth century chemistry," ibid., 99–109, and Bernadette Vincent–Bensaude, "Hélène Metzger's *La chimie*: A popular treatise," ibid., 71–84. An interesting recent discussion of eighteenth century chemistry by Evan Melhado is set largely within the conceptual framework provided by Metzger. Evan Melhado, *Jacob Berzelius: The emergence of his chemical system* (Madison, 1981), 23–60.

17. See J.R.R. Christie and J.V. Golinski, "The spreading of the word: New directions in the historiography of chemistry 1600–1800," *History of science, 20* (1982), 235.

himself, according to Maurice Crosland, has become a relatively "neglected" figure, "poorly known and little studied" by historians of science.[18] The history of science is a discipline still stretched so thinly over a vast intellectual terrain that inadequate attention to prominent problems within the field perhaps requires no special explanation. There are so many unsolved problems compared to the number of historians, that we can simply say that we have not got around to them. Nevertheless, there are signs that historians are beginning to recognize the rich unexploited opportunities that eighteenth century chemistry affords us.

2. EIGHTEENTH CENTURY CHEMISTRY AS A DISCIPLINE

Given the present preoccupation of historians of science with "social factors," it is hardly surprising that there have been calls for more attention to the institutional framework within which chemistry developed in the eighteenth century.[19] One can find numerous localized studies of institutions in which chemistry was taught or practiced, but the only broadly comprehensive treatment of the social basis of eighteenth century chemistry so far to emerge is Karl Hufbauer's *The formation of the German chemical community (1720–1795)*. Basing his interpretations on a meticulous assemblage of biographical data on forty–two men whom he identifies as active chemists in Germany between 1720 and 1780, Hufbauer has shown that the public perception of chemistry changed during this period from skepticism to enthusiastic support, and that the number of professorial chairs in medical and administrative schools expanded rapidly. He describes the range of occupational activities that these chemists pursued, especially in pharmacy and mining, as well as their teaching and research interests, and asserts that by 1780 the German chemists were coalescing into a national disciplinary community. The *Chemisches Journal*, founded in 1778 by Lorenz Crell, gave expression and further cohesion to this community. Hufbauer has added a significant new dimension to our understanding of the

18. Maurice Crosland, "Lavoisier: Le laissé pour compte," *La recherche*, 14 (1983), 785–791, on 785.
19. For example, Golinski (ref. 16), 90, 95.

formation of chemistry as a scientific discipline. His treatment nevertheless reflects the dominating historiographical influence of the canonical story, in that the only substantive chemical issue he discusses is the German response to Lavoisier's "antiphlogistic" chemistry, a debate that he believes completed the formation of the German chemical community.[20]

Studies comparable to Hufbauer's for the chemists of Britain and France are much to be desired. For the British case, Arthur L. Donovan's *Philosophical chemistry in the Scottish Enlightenment*, a monograph focused mainly on William Cullen and Joseph Black in the institutional settings of Glasgow and Edinburgh,[21] and Jan Golinski's study of Peter Shaw as exemplifying the occupational niches available to, and the patronage necessary for early eighteenth century English chemists,[22] provide strategic starting points. The most critical case, however, is France. The outlines of the institutional settings that nourished French chemistry during the eighteenth century are well known. At the Jardin du roi, two chairs in chemistry installed during the seventeenth century sustained a teaching tradition through the eighteenth. An eminent lineage of pharmacist-chemists taught chemistry in private laboratories as part of the training of apothecaries, and the subject was also taught intermittently in the laboratory of the Society of Apothecaries.[23] The central foyer for investigative chemistry throughout the century was the Académie des Sciences. Roger Hahn's *The anatomy of a scientific institution* provides an essential foundation for future explorations of the shaping effects of the Academy upon the kind of chemistry

20. Karl Hufbauer, *The formation of the German chemical community (1720–1795)* (Berkeley, 1982). Hufbauer's picture must now be supplemented and qualified by the most recent work of Christoph Meinel. See his "Zur Sozialgeschichte des chemischen Hochschulfaches im 18 Jahrhundert," *Berichte zur Wissenschaftsgeschichte, 10* (1987), 147–168.

21. Arthur L. Donovan, *Philosophical chemistry in the Scottish Enlightenment* (Edinburgh, 1975).

22. Jan V. Golinski, "Peter Shaw: Chemistry and communication in Augustan England," *Ambix, 30* (1983), 19–29. See also Golinski, "Utility and audience in eighteenth century chemistry: Case studies of William Cullen and Joseph Priestley," *British journal for the history of science, 21* (1988), 1–32.

23. Charles Bedel, "L'enseignement des sciences pharmaceutiques," in René Taton, ed., *Enseignement et diffusion des sciences en France au XVIIIe siècle* (Paris, 1964), 237–257, esp. 237–246; Yves Laissus, "Le jardin du roi," ibid., 287–341, esp. 312, 319–330.

practiced by those chemists who comprised one of the most important sub-sets among the academicians.[24]

Although the great teachers and textbooks of early chemistry—from Andreas Libavius, Jean Beguin and Nicolas Lemery of the seventeenth century to Hermann Boerhaave, Guillaume-François Rouelle and Pierre-Joseph Macquer of the eighteenth—have long been featured in histories of the subject, Owen Hannaway's brilliant book, *The chemists and the word*, has recently given us a far fuller appreciation of the nature and significance of what he has called "the didactic origins of chemistry." Fixing upon the initiator of this tradition, Libavius, Hannaway has shown how this quintessential schoolteacher composed his textbook *Alchemia* in 1597, in reaction against the mystical ethos of Paracelsian chemistry, as a means to formulate chemistry methodically for pedagogical purposes. According to Hannaway:[25]

> [*Alchemia*] is the first text which conceives of chemistry as an independent and integral discipline divorced from its applications and which seeks to organize the techniques and prescriptions of the subject in such a way that they can be taught. As such, it is much more than a late sixteenth-century encyclopedia of the operations and recipes of the various chemical arts. It marks the appearance of a new scholarly discipline on the intellectual horizon—the discipline of chemistry.

Hannaway traced only briefly, in an epilogue, the impact of Libavius's *Alchemia* on textbooks over the next 200 years. Despite the originality of his interpretation, Hannaway too bowed to the convention that all earlier chemical development finds its ultimate justification in Lavoisier. He closed with the observation that "Lavoisier's chemical revolution...was not completed until it was codified in a textbook"—the *Traité élémentaire de chemie*—which Lavoisier "designed specifically to win over the minds of youth to the new chemistry."[26]

24. Roger Hahn, *The anatomy of a scientific institution: The Paris Academy of Sciences, 1666–1803* (Berkeley, 1971).

25. Owen Hannaway, *The chemists and the word: The didactic origins of chemistry* (Baltimore, 1975), 142–143.

26. Ibid., 155–156.

John Christie and Jan Golinski have extended Hannaway's analysis to give a suggestive preliminary discussion of the pedagogical characteristics of several important eighteenth century textbooks, including those of John Freind and another Newtonian chemist, of Boerhaave, Cullen, Macquer, and inevitably, Lavoisier.[27] We are, however, still far from an understanding of these eighteenth century works matching the depth with which Hannaway has treated the origins of the didactic tradition whose later stages they represent.

The theme of the emergence of chemistry as a discipline has been treated from other standpoints as well as the didactic one. Christoph Meinel, for example, has argued persuasively that an old debate over the relation between theory and practice in chemistry acquired in the late seventeenth and retained well into the eighteenth century important implications for its standing as an academic subject. Chemists sought to distance themselves not only from alchemy, but from their craft roots in metallurgy, in order to claim for chemistry the status of a science rather than a mere art. Encountering resistance from scholars who looked down on them because they worked with their hands, chemists tried to dispel these associations in numerous programmatic lectures and introductions to textbooks, by claiming that chemistry possessed the theoretical foundations of a true scientific discipline. The prominence of Newtonian ideas, affinity theory, and matter theories in their discourses can be partially explained, in Meinel's view, by their role in this campaign for scientific legitimacy. Meinel also asserts that a fundamental reorientation of the discipline took place in the mid-eighteenth century, when the theory-practice dichotomy was displaced by a distinction between pure and applied chemistry: that is, the methodological division gave way to one based on goals. Pure chemistry still required a combination of theory and practice, but was justified as an academic discipline because its objective was the understanding of nature. Applied chemistry, meanwhile, acquired a prestige of its own because it was identified with the increasingly utilitarian values of Enlightenment society.[28]

27. Christie and Golinski (ref. 17), 235–261.

28. Christoph Meinel, "De praestantia et utilitate Chemiae: Selbstdarstellung einer jungen Disziplin im Spiegel ihres programmatischen Schrifttums," *Sudhoffs Archiv, 65* (1981), 366–389; Meinel, "Theory or practice? The eighteenth-century de-

3. THE LANGUAGE OF CHEMISTRY

The terms devised by Lavoisier and his colleagues in 1787 to designate simple substances and their combinations provided a language that expressed consistently the composition of the substances in question. To the annoyance of those who opposed the reform, the new language also codified the theoretical structure of Lavoisier's chemical system, so that adoption of the one entailed acceptance of the other. The nomenclature question has provided a strategic point of intersection between the history of science and the broader intellectual milieu of the French Enlightenment, largely through the famous preliminary discourse in Lavoisier's *Traité élémentaire de chimie*. Drawing on the principles of logic of Condillac, Lavoisier stated:[29]

> [It is impossible] to isolate the nomenclature from the science and the science from the nomenclature, considering that every physical science is necessarily composed of three things: the series of facts which constitute the science; the ideas which connect them; and the words which express them. The word must give birth to the idea; the idea must denote the fact; these three are stamped from the same mold; and...consequently one cannot perfect the language without perfecting the science, nor the science without the language.

Here too the fixation of historians on the chemical revolution has obscured our view of the language of the chemistry that preceded it. Because the nomenclature of Lavoisier has endured as the foundation of the terminology of modern chemistry, it is relatively easy for us to read Lavoisier and correspondingly difficult to penetrate the unfamiliar language of earlier chemistry. Taking our cue about the faults of the earlier language from Lavoisier's criticisms of it, we readily dismiss it as disorderly and misleading. On the one hand it appears to us to have failed to represent the compositions of the substances it named, on the other, to embody erroneous theories.

The most important recent treatment of the terminology of premodern chemistry is Maurice Crosland's *Historical studies in the language of chemistry*. Crosland has described the origins from

bate on the scientific status of chemistry," *Ambix, 30* (1983), 121–132.

29. A.L. Lavoisier, *Traité élémentaire de chimie* (Paris, 1789), *1*, vi–vii.

antiquity to the mid-eighteenth century of terms for substances based on heterogenous criteria ranging from their color, consistency, or smell to their sources, modes of preparation, medicinal virtues, and metaphorical properties. The reform of this terminology did not begin with Lavoisier, Crosland shows, but grew out of repeated expressions of dissatisfaction by eighteenth century chemists with the state of their language. Prominent among them was Pierre-Joseph Macquer, who proposed in 1766, in his *Dictionnaire de chimie*, a number of specific improvements to clarify the relationships between salts of certain classes such as the vitriols. Between 1772 and 1784 the eminent Swedish chemist Torbern Bergman developed a systematic nomenclature. Although still expressing the phlogiston theory, Bergman's system was based on general principles similar to those Lavoisier espoused shortly afterward. Guyton de Morveau, who proposed another system in 1782, was directly influenced and encouraged by Bergman. Subsequently Guyton joined with Lavoisier to produce the "new" nomenclature by modifying his previous system so as to incorporate the oxygen theory of combustion in place of the phlogiston theory. Except at those strategic points at which the theoretical structure of the new chemistry impinged upon it, therefore, the new nomenclature was not revolutionary, but the continuation of an evolutionary movement made necessary by the very success of pre-Lavoisier chemistry in expanding the number of substances requiring names.[30]

Even though he describes convincingly the links between the new nomenclature and its antecedents, Crosland still portrays the pre-reform language of chemistry mainly by its shortcomings: by the inappropriateness of some of its terms; by the cases in which the same or related terms referred to unrelated substances, or in which the same substance differently prepared was designated by different terms; and by the failure of the terminology to express composition, to group analogous substances or those sharing a common constituent.[31] We are left wondering how, in the face of such chaotic circumstances, eighteenth century chemists could think clearly or systematically at all. It may be in fact that the perceived

30. Maurice P. Crosland, *Historical studies in the language of chemistry* (London, 1962), 114–192.

31. Ibid., 65–113.

opacity of the language has inhibited historians of science from examining actively the nature of eighteenth century chemical experimentation. When they do describe substances or reactions studied by eighteenth century chemists, historians regularly resort to the expedient of translating them into modern formulas and chemical equations. Perhaps this device is helpful in gaining initial access to the phenomena in question, but it is also misleading; for the language of modern chemistry incorporates the solutions to the very problems that confronted eighteenth century chemists. I have found that with some practice it is possible to learn their own language; and that one then sees that the unsystematic nature of the terms they had inherited did not prevent these chemists from thinking coherently about the compositions, relationships, and interactions of the substances these terms designated for them.

The prominence of language in the development of eighteenth century chemistry makes the texts of this period prime targets for the application of current techniques of literary analysis. A number of historians have called for analyses of discursive practice in this field; the most ambitious venture of this type to appear so far is Wilda Anderson's *Between the library and the laboratory.* From a "close reading" of certain texts of Macquer and of Lavoisier, focused on their "literary and rhetorical dimensions"—on "the logical development of the arguments, the construction of semantic fields, the use of literary devices (such as analogy, metaphor, or imagery)" and the adoption of certain literary genres, Anderson purports to identify, between the work of Macquer and of Lavoisier, a deep displacement in the very nature of scientific practice, which she summarizes as a transformation from "philosophical chemistry" to "scientific chemistry." In her interpretation, language becomes so central to the nature of chemical analysis that laboratory work is reduced to a minor appendage.[32]

The application of recent modes of literary analysis to historic scientific texts can make a valuable contribution if the analyses are informed also by a broader knowledge of the theory and practice of the science of the time. *Between the library and the laboratory* is unfortunately restricted to a close literary reading of a small group

32. Wilda C. Anderson, *Between the library and the laboratory: The language of chemistry in eighteenth-century France* (Baltimore, 1984), 2–3, et passim.

of eighteenth century texts isolated both from related antecedent texts and from the existing historical literature on eighteenth century chemistry. The conclusions reached cannot withstand scruting from a more inclusive perspective. I hope the shortcomings in Anderson's analysis will not discourage her and others from probing such texts more deeply with a more comprehensive combination of literary, conceptual, and historical interpretation.

4. THE CHEMICAL LABORATORY

Historians of eighteenth century chemistry have turned their attention to its doctrines, its pedagogy, its institutions, its language, and to redefinitions of its disciplinary boundaries. The subject about which they have written least is what the chemists actually did in their laboratories. An exception to this general neglect of the daily activity of chemistry is a booklet written a decade ago by Jon Eklund, entitled *The incompleat chymist.* Deploring the tendency of historians to "ignore the details of practice," Eklund argues that chemistry was pre-eminently a science not of theory but of operations. The most generally accepted definition of chemistry in the seventeenth and eighteenth century was that it was the art of separating the substances that enter into the composition of bodies, and recomposing the same bodies from these same substances. Eklund portrays "the eighteenth century chemist in his laboratory," describing his equipment, the procedures customarily employed, the reagents ordinarily on hand, the means available for obtaining pure substances, and so on. He then gives detailed examples of representative research operations carried out by three eighteenth century chemists. In order to encourage historians to read the technical literature of the period, he provides a convenient short dictionary of the chemical terms commonly used.[33]

Chemical laboratories of the mid-eighteenth century differed little from their counterparts stretching back for at least a century and a

33. Jon Eklund, *The incompleat chymist: Being an essay on the eighteenth century chemist in his laboratory, with a dictionary of obsolete chemical terms of the period* (Washington, D.C., 1975).

half. The repertoire of apparatus in use and the classes of operations performed in the laboratories of the 1750s were similar to those described in Andreas Libavius's *Alchemia* in 1597; and Libavius himself assembled for his methodological presentation accounts of procedures already long in use. In the mid-eighteenth as in the late-sixteenth century, the laboratory normally consisted of a large room, dominated by several furnaces of varied designs adapted to different types of operation and by an assortment of vessels that were assembled in various combinations to perform distillations and sublimations. There was also an array of open vessels, utensils and materials for operations such as filtrations, evaporations, macerations, calcinations, coagulations, precipitations, and other subtle variations of these basic means for separating or combining substances. At the end as at the beginning of this extended epoch, fire dominated the scene, visually and operationally, as the most powerful agent available to effect chemical change, and chemists practiced a highly refined craft skill to control a charcoal or wood fire in order to adapt it to the particular conditions of heat necessary to carry out a myriad of special processes.[34]

Until late in the eighteenth century no major technological changes altered the character of the chemical laboratory as a material or social setting. The advent of accurate, reliable thermometers early in the century had little impact on standard practices, because few chemists adopted them then. Physical conditions in laboratories were subject to little more artificial control in 1750 than in 1600. In 1766, in a description of the typical chemical laboratory in his dictionary of chemistry, Macquer pointed out that in deciding between the ground floor and an upper story for a laboratory one had to choose between easy access, storage, and water supply in the former, but with poor ventilation and moisture that ruined reagents and caused labels to fall off of bottles; or the drier conditions, better light and ventilation of the latter, compensating for the difficulties of access and supply. Interiors of buildings were so incompletely isolated from exterior conditions that the success of chemical operations often depended upon the season during which they were undertaken.[35]

34. Friedemann Rex, ed., *Die Alchemie des Andreas Libavius: Ein Lehrbuch der Chemie aus dem Jahre 1597* (Weinheim, 1964), 1–116.

35. [P.-J. Macquer], *Dictionnaire de chymie* (Paris, 1766), 2, 1–11.

The stability of the chemical laboratory through the two centuries in which chemistry emerged as a scientific discipline suggests that it would be fruitful to treat this institution from the perspective that Fernand Braudel has called the "longue durée." For a very long time this prototypical scientific laboratory provided a milieu that was "quasi immobile," as Braudel termed the far greater milieu that the Mediterranean provided for the cultures that endured within it.[36] The physical layout of the chemical laboratory both offered the opportunities for and fixed the limits on what chemists could do. As a material structure, the laboratory also shaped the social position of the chemist, holding his vocation in an ambivalent status between that of a craft and a science. If chemists sought through their theoretical doctrines, their pedagogy, and their literary activities to attain the standing of natural philosopher, they were nevertheless viewed in less august ways because of the environments in which they worked. The fact that they operated with charcoal and furnaces, doing outwardly very much what metallurgists, assayers, apothecaries, and even bakers did, made it very difficult for them to escape the image of "sooty empirics."

Over its longue durée, the chemical laboratory was a stable, but not a static institution. Braudel has shown in his monumental study of the Mediterranean that changes in the material conditions pertinent to the cultures existing there took place at an almost imperceptible pace. On a far smaller scale, the chemical laboratory too evolved very gradually, at a pace measured not in years but in generations of chemists. Operations were refined, the designs of furnaces improved. New reagents were added, and chemical indicators supplemented the basic tests of color, odor, and taste. Most important, the purposes for which the operations were performed were slowly transformed. The medical objective of separating active "essences" of substances from the grosser matter in which they inhered, or of enhancing their qualities, and the metallurgical objective of preparing metals from ores were supplemented and eventually overshadowed by the chemical goal of decomposing and recomposing substances in order to determine the nature of their

36. Fernand Braudel, *La Méditerranée et la monde méditerranéen à l' époque de Philippe II*, 5th ed. (Paris, 1982), *1*, 13. See also, Braudel, "Histoire et sciences sociales: La longue durée," in *Ecrits sur l'histoire* (Paris, 1969), 41–83.

constituents. Conceptions of the nature of these constituents themselves were also gradually transformed. Within the same material setting, with much the same apparatus, using mainly the same operative procedures and limited by the same physical constraints as their forebearers of a century and a half earlier, the chemists of the mid-eighteenth century were engaged in a different enterprise.

Up until the middle of the seventeenth century, the gradual changes to which I have alluded occurred mainly within craft and pedagogical contexts. Conceptual and technical changes resulted from efforts to improve practical chemical processes and to teach the art more effectively. The events constituting what we call the scientific revolution, however, drew chemistry into a new context as a branch of natural philosophy. Best known in this regard, especially through the historical treatment of Marie Boas, is the achievement of Robert Boyle in England.[37] In Boyle's hands the traditional operations of the chemical laboratory were redirected toward the experimental exploration of material processes. Less known, but in my opinion of even greater long-term significance, was the incorporation of chemistry as a major activity of the French Academy of Sciences from the time of its founding in 1666. There chemistry became one of several areas in which the academicians carried on sustained research. By the early eighteenth century, the Academy was the most important center for systematic chemical experimentation.

This association of the chemical laboratory with an institutional ethos that placed special value on communal and then individual research, on novel discoveries and conceptual originality, transformed chemistry not only into an experimental science, but into an increasingly dynamic investigative enterprise. In the next two lectures I discuss two areas of investigation, independent of those canonical events afterward seen as steps toward the chemical revolution, in which eighteenth century chemists achieved conspicuous success.

37. Marie Boas, *Robert Boyle and seventeenth-century chemistry* (Cambridge, 1958).

ABBILDUNGEN AUS

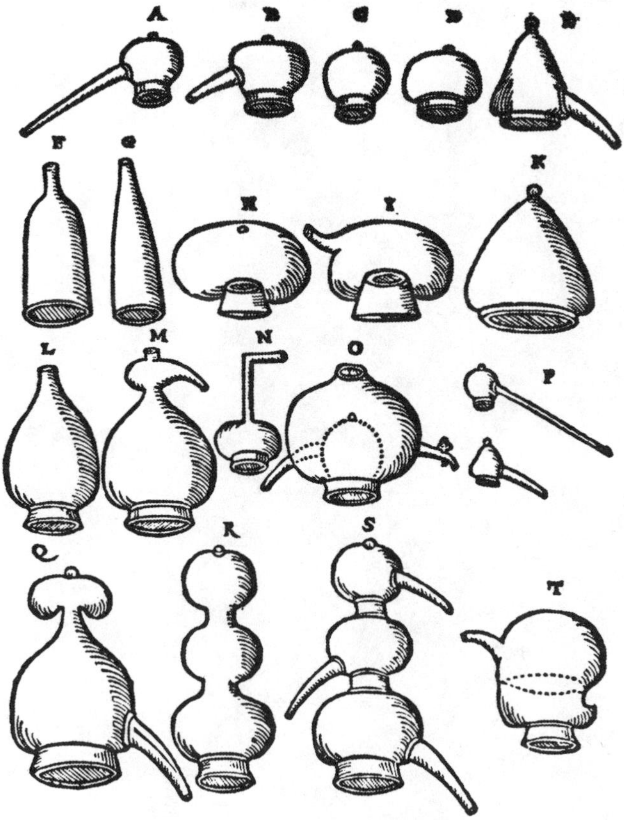

Abb. 107 ⟨173⟩. [Aufnahme- und Überleitegefäße ⟨accipientia, tradentia⟩].

FIG. 1 Various forms of "receiving and transmission vessels," to be placed on top of distillation vessels; A, head with long nose; C, blind head with hole in top; F, sublimation head; O, head with cooler; S, three-part head. From Friedemann (ref. 34).

FIG. 2 An eighteenth century French chemical laboratory, as depicted in *Recueil des planches pour la nouvelle édition du dictionnaire raisonné des sciences, des arts et des métiers* (Geneva, 1779), "chemie," following p. 11.

FIG. 2 Among the numbered items of apparatus are: **3**, aludels for obtaining the spirit of sulfur according to the method of Stahl; **4**, a recipient or balon; **5**, **6**, retorts; **8**, matras (distillation vessel) with head; **9**, **10**, funnels; **12**, *jumeaux* or twin distillation flasks; **13**, pelican; **14**, an alembic with cucurbite (forms of distillation apparatus); **15**, Venel's vessel for obtaining essential oils; **16**, *Matras de rencontre* (another form of distillation apparatus); **17**, vessels for sublimation of *flowers of benzoin;* **18**, apparatus for measuring the quantity of air given off by a fermenting body (an early visible sign of a pneumatic apparatus added to the traditional equipment of a chemical laboratory); **19**, **20**, mantle for the chimney (the large size of the mantle, under which the furnaces are placed, indicates the importance of ventilating the fumes produced in a chemical laboratory); **21**, bellows for the forge; **22**, *bain marie* for a glass cucurbite (a water bath placed under distillation flask to maintain even, moderate heat); **23**, double serpentine in a tub; **24**, **25**, lower (cucurbite) and upper (head) sections of copper distillation vessel; **27**, athanor (a digesting furnace); **28**, **29**, *matras* and *tower* heated by athanor; **31**, laboratory table; **32**, glass vessels containing solutions of metals; **34**, assay furnace; **35**, filtration funnels; **36**, support table for funnels; **43**, small furnace; **45**, apparatus for producing a *clyssus* (see Lecture 4); **46**, forge; **47**, water basin under water fountain; **49**, water barrel. The numbered individuals are: **30**, physician consulting with the chemist; **33**, the chemist; **44**, a chemist producing a clyssus; **26**, laboratory assistant carrying charcoal; **48**, laboratory assistant washing vessels; **50**, another laboratory assistant

LIBAVIUS (COMMENTARII I, S. 173/4)

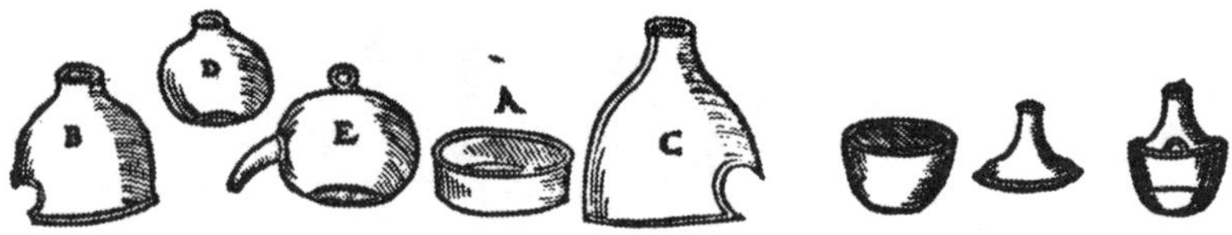

Abb. 108/109 ⟨173/4⟩. [Spezielle Tiegelaufsätze].

FIG. 3 Above, crucibles; below, various forms of recipient for collecting distillation products. Source as for fig. 1.

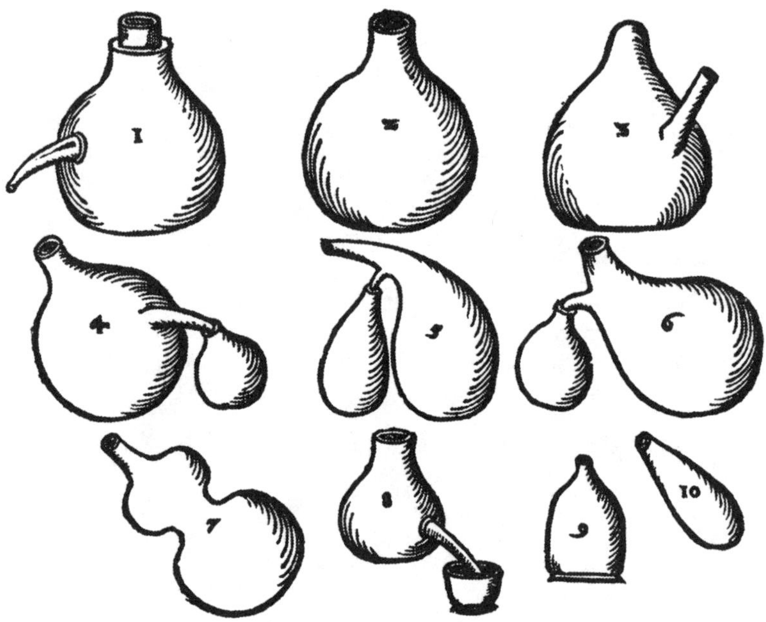

Abb. 110 ⟨174⟩. [Sammelgefäße ⟨collectoria⟩].

Abb. 152 ⟨310⟩.
[Bad mit Stufa sicca].

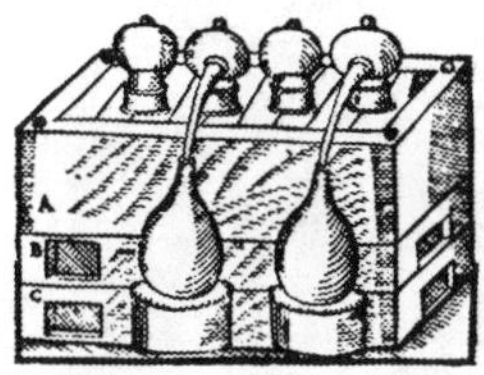

Abb. 153 ⟨310⟩.
[Destillation im Aschen- oder Sandbad ⟨cinerarium vel arenarium⟩].

Abb. 154 ⟨311⟩.
[Destillation über eine Schlangenwindung ⟨destillatorium anguinum⟩].

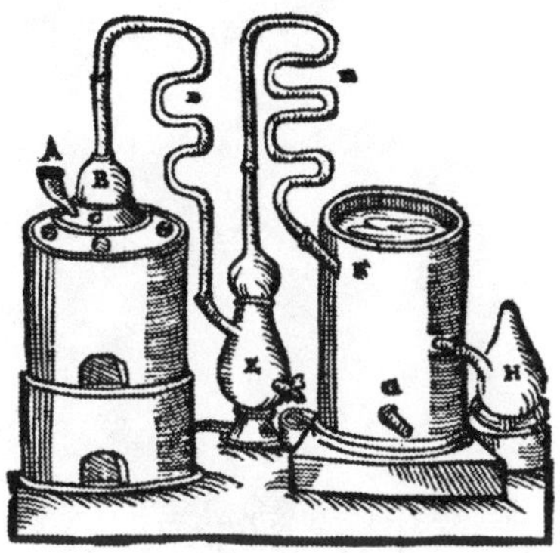

Abb. 155 ⟨311⟩. [Destillation über mehrere Windungen]
[vgl. PORTA 414].

FIG. 4 Various arrangements of distillation vessels in place in furnaces: bf 152, distilllation vessel in water bath; bf 153, distillation vessels in sand bath; bf 154, distillation through winding tubes passing through coolant water. Source as for fig. 1.

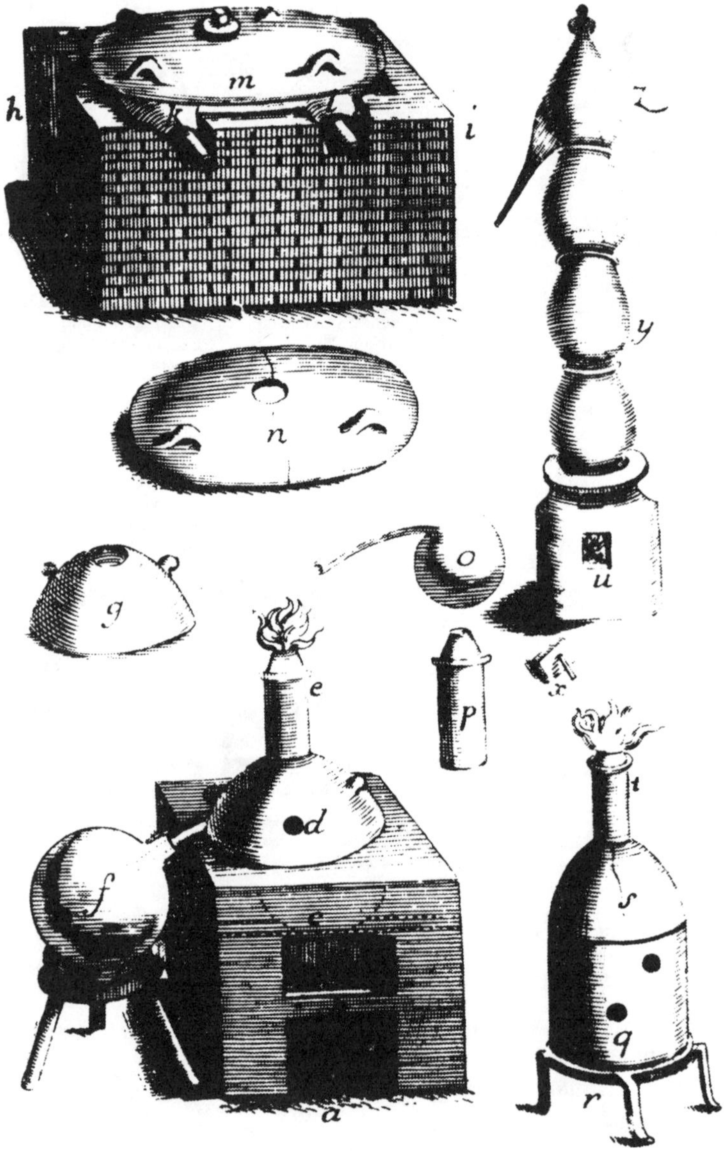

FIG. 5 h, k, m, l, reverboratory furnace containing two retorts without recipients; a, b, c, d, e, f, reverboratory furnace with a single retort and recipient. From Nicolas Lemery, *Cours de chemie*, 9th ed. (Paris, 1701), 42–47.

FIG. 6 q, matras (form of distillation vessel with long vertical neck); k, l, m, n, o, distillation apparatus with large copper cucurbite (1), and copper tube passing to recipient through barrel filled with water; r, s, matras with blind head, for sublimations; a, b, c, d, e, f, g, large reverboratory furnace with six retorts. Source as for fig. 5.

FIG. 7 k, l, m, n, o, p, q, distillation apparatus with large tin-plated copper cucurbite (n) and water-cooled head (o); b, c, d, g, e, h, same but with long tin "serpentine" (d); at bottom, two small portable furnaces. Source as for fig. 5.

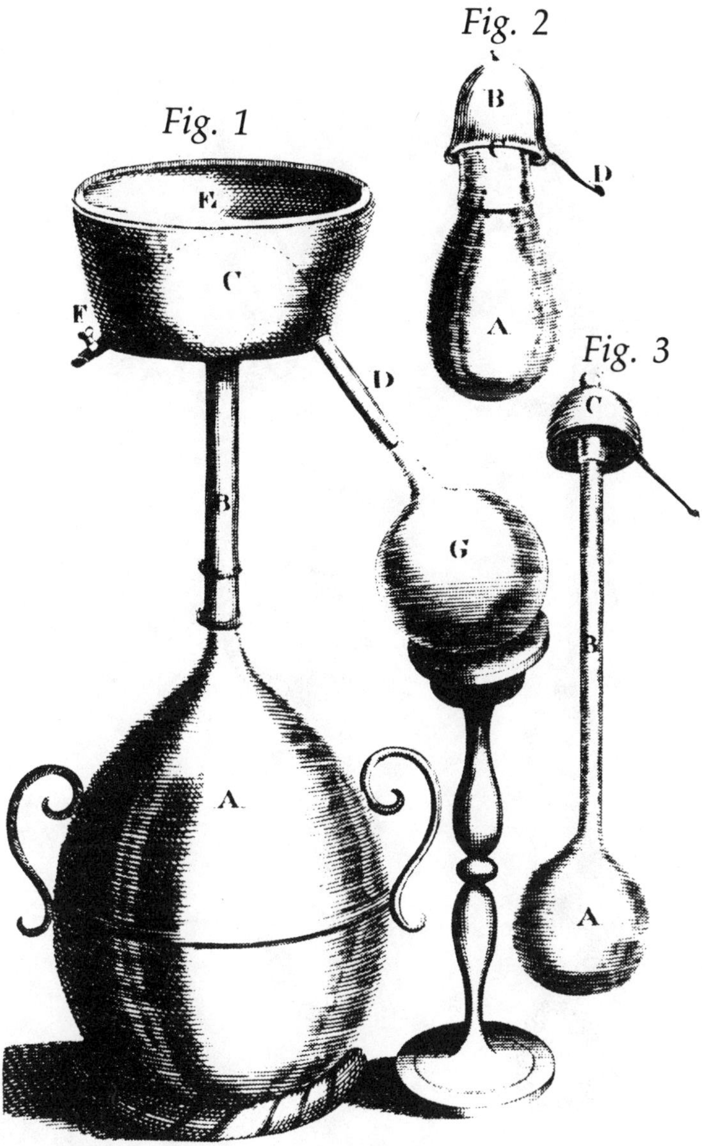

FIG. 8 *Fig. 1*, a copper alembic (distillation apparatus): A, the cucurbite or body; B, the neck; C, the head; D, the beak, nose, or spout; E, the refrigeratory (filled with water); G, the receiver. *Fig. 2*, a glass alembic. *Fig. 3*, a long-necked glass alembic. From P.J. Macquer, *Elements of the theory and practice of chymistry*, 5th ed. (Edinburgh, 1771), after p. 620.

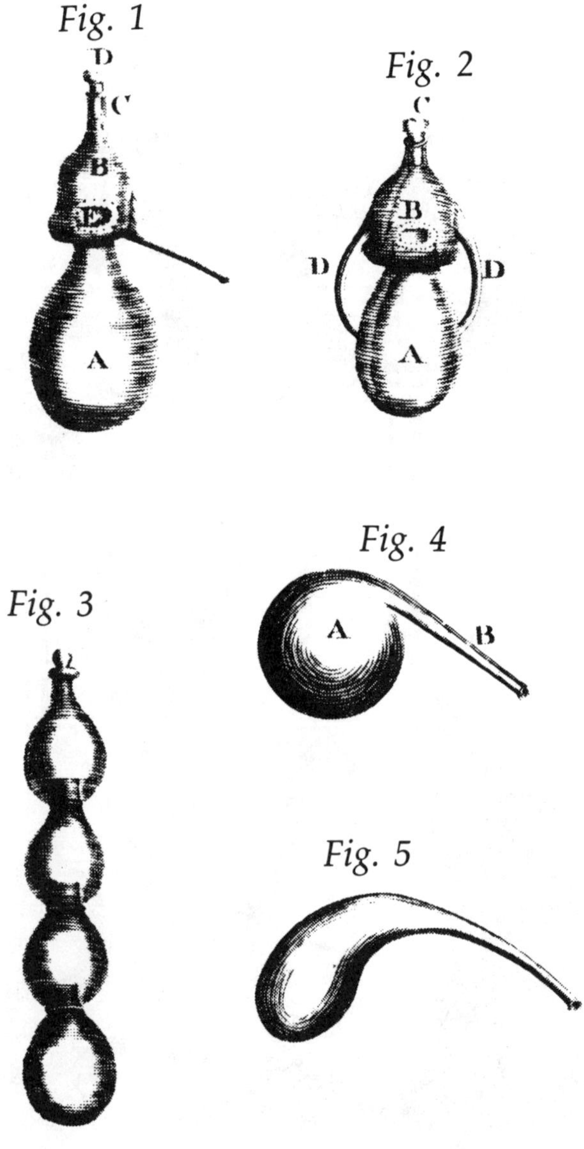

FIG. 9 *Fig. 1,* one piece glass alembic. *Fig. 2,* a pelican. *Fig. 3,* a row of aludels (multiple head for distillation apparatus, used for sublimations). *Figs. 4–5,* retorts (used for "lateral" distillation of less volatile substances). Source as for fig. 8.

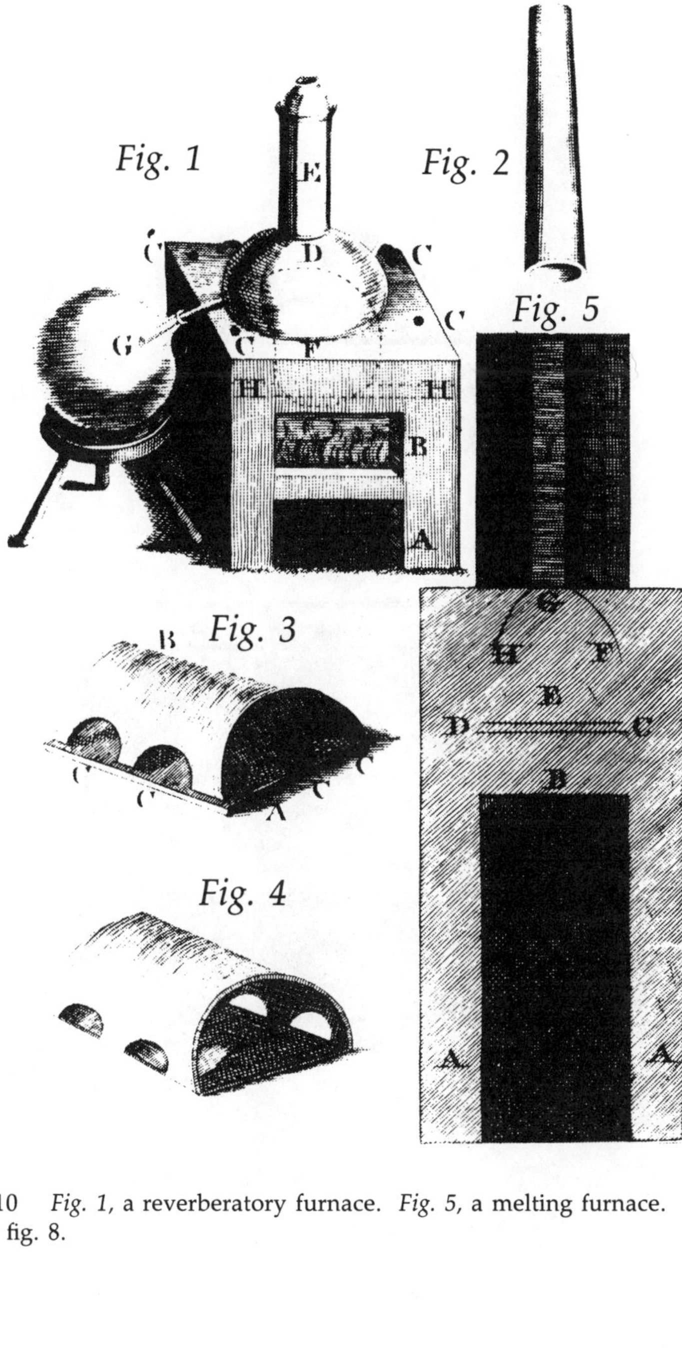

FIG. 10 *Fig. 1*, a reverberatory furnace. *Fig. 5*, a melting furnace. Source as for fig. 8.

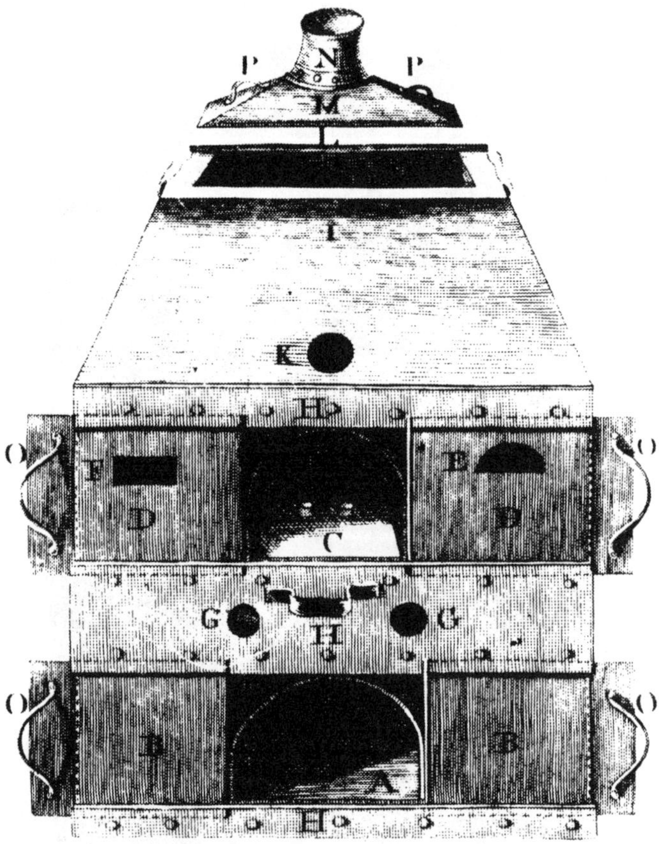

FIG. 11 A cupelling furnace. Source as for fig. 8.

2 THE CHEMISTRY OF SALTS

1. INVESTIGATIONS IN THE PARIS ACADEMY OF SCIENCES

During the first decade of the eighteenth century, there were assembled in the *Académie royale des sciences* a group of the ablest chemists in Europe. In their ages they represented three generations. The senior figure among them was in the late stages of a prominent career. Beginning as an apothecary, Nicolas Lemery had opened in Paris in 1672 a private course in chemistry that attracted students from all over the continent. He had also written a textbook, *Cours de chymie,* which described and interpreted chemical operations so lucidly that it supplanted all earlier works of its type, and was in its eighth edition by the time Lemery entered the Academy in 1699. The most conservative member of the group was Simon Boulduc, formerly apothecary to the Queen dowager of Spain and demonstrator in chemistry at the Jardin du roi, who became an academician in 1694. Boulduc concerned himself mainly with the traditional objective of improving drugs. As we shall see in the next lecture, however, he proved to be an innovative investigator. The dominant chemist within this group was Wilhelm Homberg. A man of broad scientific interests who had travelled widely and studied with scientists in all of the major European countries before settling in Paris, Homberg was appointed to the Academy in 1691.[38]

38. "Eloge de M. Lemery," *HAS,* 1715 [publ. 1718], 73–82; "Liste de Messieurs de l'Académie Royale des Sciences," ibid., *2* (Paris, 1733), 373–374; Bernard de Fontenelle, "Éloge de Homberg," in Fontenelle, *Oeuvres* (Paris, 1825), *1,* 308–316.

The youngest generation of the chemists in the Academy included Lemery's son Louis, appointed at the same time as his father; Boulduc's son Gilles; Etienne-François Geoffroy, a pharmacist and physician who had entered the Academy in 1699 at the age of 27 as a "student" of Homberg and was promoted quickly to an associate Academician; and Geoffroy's brother Claude-Joseph, thirteen years younger, also a pharmacist, who was elected in 1706.[39]

By the end of the seventeenth century, the ambitious communal projects initiated three decades earlier by the first generation academicians had become routine and uninspired. Recognizing the benefits of individual scientific enterprise, the administrators of the Academy provided in the revised regulations of 1699 that each academician was thenceforth expected to commit himself annually to some particular topic of investigation, and to present discussions of his results at regular intervals to the assembled society.[40] The new ethos rewarded discoveries of new phenomena, innovative solutions to recognized problems within a shared specialty, and other indications of originality. As Etienne Geoffroy put it in a letter to the Secretary of the Royal Society in 1709, the academicians were obligated "to contribute through their works everything in their power to advance the sciences and to distinguish the Academy, and to communicate to the company from time to time what they have learned that is new in the sciences."[41] Beginning in 1700, the Academy published an annual volume of memoirs comprised mainly of the texts of these progress reports of the individual academicians to their colleagues. As I have discussed elsewhere, this form of publication consolidated the literary genre of the scientific research paper devoted not to the didactic treatment of a broad subject matter, but to a presentation of the author's most recent contribution to a limited set of problems.[42] The chemists of

39. "Liste de Messieurs" (ref. 38), 377–393; "Éloge de Geoffroy," in Fontenelle, *Oeuvres, 1,* 281–289; W.A. Smeaton, "Geoffroy, Claude Joseph," *Dictionary of scientific biography, 5,* 351–352, and "Geoffroy, Etienne François," ibid., 352–354.

40. *HAS,* 1699, 2nd ed. [Paris, 1718], 1–16.

41. I. Bernard Cohen, "Isaac Newton, Hans Sloane and the Académie Royale des Sciences," in I.B. Cohen and R. Taton, ed., *L'aventure de la science,* 2 vols. (Paris, 1964), *1,* 61–116, on 90.

42. Frederic L. Holmes, "Argument and narrative in scientific writing: The *Académie des Sciences* and chemical accounting," in Peter Dear, ed., *The paper laboratory: Textual strategies, literary genres and disciplinary traditions in the history of science* (Philadelphia, forthcoming).

the Academy responded with a vigorous flow of memoirs reporting major and minor additions to what they saw as a lively field of experimental enquiry.

For two decades Homberg was the most productive author of such chemical memoirs. Most of them were reports of his own experimental efforts and the immediate conclusions he could draw from them. In 1702, however, he undertook a series of broader essays bringing his prior investigative experience to bear on current general ideas about chemical composition. The first article was on salts. There were, he said, three types of simple salts: fixed (or lixivial) salt, and two volatile salts, the urinous and the acid. These were not novel identifications. They represented three classes—more commonly known as acid, fixed alkali, and volatile alkali—that could be distinguished by chemical indicator tests introduced forty years earlier by Robert Boyle. In addition, however, Homberg described a third class of salts that had not previously been identified so generally or so distinctly:[43]

> The acids joined to the fixed salts compose mixt salts, or middle salts, according to the nature of the acids which are engaged in them; for example, spirit of nitre joined to salt of tartar produces true saltpeter, spirit of salt joined to salt of tartar produces common salt, spirit of vitriol joined to salt of tartar produces true vitriol, but without metal, etc., which are all middle salts, that is, in part fixed and in part volatile, because the two salts composing them are and remain the one fixed and the other volatile.

Homberg identified in these examples the three salts that would be formed by joining one of the two known alkalis—that is, "fixed alkali," called here salt of tartar after its most common source—in turn with each of the three common mineral acids, here still called by the more general name "spirits" of their respective three sources. As the passage suggests, Homberg called the combinations "middle" salts because in each case one of the components was "fixed" and the other "volatile." These two adjectives meant respectively that in a distillation the latter was raised, or "flew up" into the receiver of the distillation apparatus, while the former remained fixed in the distillation flask.

43. Wilhelm Homberg, "Essays de chimie: Article premier. Des principes de la chimie en general," *HAS*, 1702, 2nd ed. (Paris, 1720), 33–52, on 40–41.

A second group of middle salts, according to Homberg, was formed by combinations of each of the same three mineral acids with an alkaline earth, such as is found in lime. A third group resulted from the joining of these acids with each of the metals. In addition to the middle salts there was a whole class of sal ammoniacs. Common sal ammoniac was a combination of spirit of salt (or marine acid) with volatile alkali. The other sal ammoniacs were made by substituting "in place of the spirit of salt whatever acid mineral salt one wished." Homberg had himself discovered in the previous year another special sal ammoniac containing a vegetable acid in place of the spirit of salt.[44] These sal ammoniacs differed from middle salts in that both of their constituents were volatile.

Homberg's classification of middle salts and sal ammoniacs represented the systematic generalization of a mode of interpreting the composition of substances and chemical operations that had been applied occasionally to specific cases as early as a century before, and had been very slowly generalized over the long intervening period. At the beginning of the previous century, the dominant view had been that all substances are composed of the same few elements or principles. The most common of a variety of schemes, a compromise between the Aristotelian four elements and the three principles postulated by Paracelsus, was that all matter is composed of salt, sulfur, mercury, earth, and water.[45] Despite the fact that most mainstream seventeenth century chemists no longer inhabited the philosophical world of either Paracelsus or Aristotle, they long strove to identify the products of their distillations and other decomposition processes with these five principles. Concurrently, they found it possible to interpret the composition of certain mineral substances more pragmatically in terms of tangible constituents into which they could decompose them and from which they could sometimes recompose them. During the course of the century, vitriols came to be described with increasing clarity as combinations of metals with spirit of vitriol. By the middle of the century both Rudolph Glauber and Robert Boyle had decomposed nitre into salt of tartar and spirit of nitre, had recomposed nitre from the

44. Ibid., 41–42.

45. See R. Hooykaas, "Die Elementenlehre der Iatrochemiker," *Janus, 41* (1937), 1–28.

acid and the alkali, and understood the nitre as composed of these two constituents.[46] By the end of the century these operational interpretations of composition had gained a practical ascendancy over the older conceptions based on principles and elements, because the latter, unlike the former, provided little guidance in differentiating the specific substances encountered in experimental practice.

Although pragmatic, the concept of the middle salt was far from a simple or direct empirical generalization. To arrive at it, chemists had to accept a proposition that was very difficult for them to comprehend; that is, that a substance could still be present within a combination even though its characteristic properties had disappeared. The difficulty was heightened by the fact that the traditional ideas of composition stood on the assumption that each of the properties manifested by any specific substance—such as solidity, volatility, fixity, or inflammability—were embodied in one of the elements or principles constituting it. As the reactions of acids and alkalis came to the center of attention in the middle of the seventeenth century, the most striking feature of the operation to chemists was that these substances were of opposite nature, that they reacted violently together, and that afterward neither the properties of the acid nor of the alkali were evident. Chemists conceived of the process therefore as a "combat" in which the alkali and the acid destroyed one another. Nicolas Lemery explained this combat in his textbook in terms of the currently fashionable mechanical philosophy, by imagining the acids to be composed of sharply pointed particles. In the violent interaction the points broke off, so that the acid lost "all of its acidity."[47] The application of color indicators to identify acids and alkalis fit into the view that neither the acid nor the alkali were present after the two had reacted together; for the resulting salt did not turn the indicator to the characteristic color identifying an acid or an alkali.

Just as corpuscular images had helped mechanistically oriented chemists to understand acid–alkali reactions as destructive combats, so Homberg resorted in 1702 to an altered corpuscular image to

46. John Rudolph Glauber, *The works*, trans. C. Packe (London, 1689), pt. 2, 98; Robert Boyle, *Certain physiological essays* (London, 1661), 107–114.

47. Nicolas Lemery, *Cours de chymie*, 5th ed. rev. (Paris, 1683), 28.

help him picture his new conception of a middle salt:[48]

> Acid salts are ordinarily considered as antagonists of these alkalis, because their mixture hardly ever takes place without a great ebullition and effervescence; but one could say more reasonably that [there is]...a very well–fitted joining of two materials....They are also comparable, the one to sheaths and the other to points appropriate for being inserted into these sheaths.

The corpuscular image thus made it plausible for Homberg that the acid could exist unchanged in the combination even though its properties were not evident. The explanation was in a sense a metaphor. The sharp taste of an acid is masked in a middle salt just as a sharp point is covered by a sheath.

Mechanistic images faded from discussions of specific chemical processes after the first decade of the century. It remained difficult to understand how constituents of a substance could exist intact within it without manifesting their characteristic properties, and eighteenth century chemists continued to try to explain properties common to a class of substance by invoking an underlying principle contained in all of them. Nevertheless, at an operational level the conception of a class of middle salts (or as they were later more often called, neutral salts) composed of any combination of the known acids with any of the known alkalis, alkaline earths, or metals, became increasingly dominant in eighteenth century chemistry. The reason was that it served as a powerful organizing principle for experimental investigation.

By 1718 Homberg and the elder Lemery had died, and Etienne Geoffroy had risen to prominence, not only in the Academy, but as Professor of Medicine at the *College de France* and Professor of Chemistry at the Jardin du roi. In that year Geoffroy published in the memoirs of the Academy a short article entitled, "Table of the different *rapports* observed between different substances in chemistry." Observing that different chemical substances have propensities to join together in preference to others, he claimed that one could deduce from many observations the general proposition that "every time that two substances which have some disposition to join together, one with the other, are found together, if a third one

48. Homberg (ref. 43), 41.

arrives which has more *rapport* with one of the two, it unites with it, causing the other to be let go."[49] Accompanying the article was a table of sixteen columns, each column containing a list of substances—represented by alchemical symbols—placed vertically corresponding to the order in which they replace one another in combination with the substance at the head of the table.

Geoffroy's table has become celebrated for the introduction of affinity theory into eighteenth century chemistry. Some historians have argued that *rapports* was a thinly disguised concept of "attractions," and that Geoffroy was bringing into French chemistry the view expressed so forcefully in the Queries of Newton's *Optics,* that substances come together or separate according to the degree of the attractive forces exerted between them at very short distances. Support for this interpretation has been drawn from the fact that Geoffroy corresponded regularly with the Royal Society and that he had in 1706 presented an extended abstract of the *Optics* at a series of meetings of the Academy.[50] Others have argued, however, that Geoffroy could not have intended such a meaning, because his mechanistic ideas about the particles of matter were more akin to those of Nicolas Lemery or Descartes than to the views of Newton.[51] Jean-Claude Guédon has argued that Geoffroy's tables had little to do with Newtonian attractions, but introduced a radical rupture into chemistry, transforming it from a science of materials to one of reactions.[52] Crosland has maintained, on the other hand—as I pointed out in the previous lecture—that the primary importance of the affinity tables was simply as a classification of experimental evidence, "summarizing chemical reactions and predicting others."

Overlooked in most of these historical discussions of Geoffroy's table is the particular character of the chemical substances and changes summarized in it. The first eight of its sixteen columns were filled mainly by the three mineral acids, fixed and volatile alkali, "absorbent earth," and "metallic substances." The substances that would be formed by their combinations were middle

49. E.F. Geoffroy, "Des differens rapports observés en chimie entre differentes substances," *HAS,* 1718 [publ. 1719], 202–212.

50. Cohen (ref. 41), 77–116; Thackray (ref. 3), 89–92.

51. Smeaton (ref. 39), 354.

52. Jean-Claude Guédon, "Remarques sur la revolution chimique," unpubl. ms., Institut d'Histoire et de Sociopolitique des Sciences (Montreal, 1976), 1–9.

salts. In other words, Geoffroy provided a dynamic interpretation of the same combinations and constituents that made up his former mentor Wilhelm Homberg's classification of middle salts and sal ammoniacs.

Geoffroy stated that he had derived his table "as much from the experiments and observations of other chemists as from my own." He announced his intention to publish in future articles the experiments on which he had based the order of substances; but he did not do so, and it is therefore difficult to determine how much of the table was based on his work and how much on a review of the existing literature. The substances and most of their combinations were familiar; but, as he remarked, "the order of the *rapports* is little known."[53] It may have required an extensive investigative program to sort them out. In any case, the table was a nodal point in the continuing evolution of a pragmatic chemistry of operations oriented around the concept of the middle salt.

In the text Geoffroy summarized verbally the order of the chemical operations indicated so compactly in the table. He then provided a detailed example of the manner in which the tables enabled one to explain what happened in a complex chemical process, in this case the preparation of corrosive sublimate. He interpreted several different methods that had been used to produce this substance, and showed through the tables not only what acids and bases combined and separated in the process, but how this understanding could be used to simplify the procedures.[54]

The ordinary method began with calcined vitriol of iron, marine salt, and mercury dissolved in spirit of nitre evaporated to dryness. The dry mixture was placed in a retort, and the dried salt was obtained by sublimation (the name corrosive sublimate, which chemists knew to be composed of mercury and the acid of marine salt, referred to its mode of preparation and its most striking property). Some chemists, Geoffroy reported, made corrosive sublimate in a different way, "by dissolving mercury in spirit of nitre and pouring this solution into a solution of marine salt in water, after which a white precipitate formed." The precipitate was separated by decantation, dried, and distilled to obtain the product in the

53. Geoffroy (ref. 49), 202–203, 212.
54. Ibid., 202–212.

form of the sublimate.[55]

The juxtaposition of these two methods is illuminating. The first typified what historians have regarded as a traditional procedure. Substances were mixed dry, and the desired product separated out by means of fire in a distillation apparatus. The second relied on getting the substances into solution and separating the product by its relative insolubility. The second, or "wet" way, is thought to have come into widespread use only around the middle of the eighteenth century. Geoffroy's example suggests, and I believe that the prior literature would support, that both approaches had long been in the chemist's repertoire. The shift in emphasis toward solution methods from the early eighteenth century onward was probably driven to a large degree by the effectiveness of these methods in managing the emerging chemistry of middle salts.

Homberg's classification of middle salts and Geoffroy's table provided conceptual frameworks to organize and clarify knowledge of chemical substances and operations that had been accumulating over the preceding century. Geoffroy's table accounted for thirty-two different combinations of the three mineral acids and spirit of vinegar with fixed and volatile alkali, absorbent earth, iron, copper, lead, mercury, silver, tin, regulus of antimony, and sulfur. During the decades after 1720 this conceptual structure began to expand, as chemists applied it and the solution methods with growing power to discover new species within each of the categories of acid, alkali, alkaline earth, and metal.

There had long been evidence that the alkali separable from sea salt was somewhat different from that derived from salt of tartar and other plant substances. For example, the salts formed by vitriolic acid with alkalis from these two sources were recognized as distinct. Early eighteenth century chemists nevertheless continued to identify these alkalis interchangeably under the common designation fixed alkali.

Henri-Louis Duhamel du Monceau, a versatile scientist who had become adjunct chemist of the Academy in 1728 but who carried on research on the cultivation and physiology of plants as well as on chemical problems, undertook in 1736 to resolve the ambiguity concerning the fixed alkalis.[56] His elegant approach to the problem

55. Ibid., 206–212.

56. Claude Viel, "Duhamel du Monceau, naturaliste, physicien et chimiste," *Revue d'histoire des sciences*, 38 (1985), 55–71.

illustrates the efficacy of the investigative strategies becoming available to chemists. The difficulty in identifying the base contained in sea salt lay, he believed, in the fact that it had never been obtained separated from all acids so that its properties could be directly examined. The salt composing it was not decomposable by distillation, so that the only way to remove the acid was to displace it by another mineral acid, which then remained combined with the alkali. After attempting in several ways unsuccessfully to remove the marine acid, Duhamel achieved his goal by displacing it with vitriolic acid. There was no direct way then to remove the vitriolic acid. By calcining the vitriolic salt with charcoal, however, he was able to reduce the acid to sulfur, or so he assumed, by means of the inflammable matter in the charcoal. The sulfur being more readily separable than vitriolic acid, he could displace it from the alkali with the weak acid of vinegar. The acid of vinegar was removable by distillation, leaving a residue that he identified through the usual indicator tests as a fixed alkali.[57]

Having isolated the fixed alkali, Duhamel showed that it differed from the alkali derived from tartar in that it did not liquify in air but formed a powder. Moreover, with each of the three mineral acids and acid of vinegar it formed a middle salt that resembled but was distinguishable from the corresponding salt formed by salt of tartar. The salt formed with marine acid was identical to common salt, so that Duhamel could claim that he had proven the composition of the latter by analysis and synthesis.[58] The new alkali soon became known as fixed mineral alkali, the more common one as fixed vegetable alkali. Duhamel's investigation not only identified a new alkali, but distinguished at least four middle salts of the alkali from those of the other fixed alkali. Although he did not explicitly refer to Geoffroy's table, the displacement reactions that Duhamel employed in order to attain these results followed the order predicted by the table. His investigation thus both displayed the effectiveness of the new chemistry of salts and significantly expanded its scope.

57. Duhamel, "Sur la base du sel marin," *HAS*, 1736 [publ. 1739], 215–232, on 215–226.

58. Ibid., 226–230.

The substance identified in Geoffroy's table as "absorbent earth" referred to a group of similar materials such as limestone, coral, chalk, and powdered sea shells that were insoluble in water but effervesced with the mineral acids in a manner similar to the alkalis. They were therefore sometimes called alkaline earths. Each of these materials when calcined yielded lime, which dissolved in water with ebullition. Left standing, limewater deposited an insoluble whitish crust. It was difficult to reconcile the properties of these substances with the traditional sharp distinction between earths as inert, insoluble substances and salts as active substances characteristically dissolving in water. Chemists had long debated whether there might be contained in these earths a salt that would account for the behavior resembling that of an alkali.[59]

A further complication developed when it appeared that the combination of an alkaline earth with an acid might itself be barely soluble. In 1726 Gilles Boulduc, one of the younger chemists who had joined the Academy at the beginning of the century but had been generally less active in research than his colleagues, found when he evaporated a particular mineral water that a substance fell to the bottom in the form of regular crystals. This mode of formation seemed to Boulduc sufficient to demonstrate that the material was a middle salt. By melting the crystals with salt of tartar (that is ordinary fixed alkali) and dissolving the product in water, he precipitated an earth that could be redissolved in acids. When he evaporated the remaining solution he obtained crystals of vitriolated tartar (a middle salt understood to be composed of fixed alkali and acid of vitriol). The first crystalline material, he therefore concluded, was a salt containing alkaline earth and acid of vitriol. He called it selenite because of its similarity to certain crystalline minerals by that name. Besides identifying a new middle salt, Boulduc also stretched the definition of salts. In the memoir he published on his discovery he argued that solubility was not essential to the character of salts; for those who might not accept that argument, he added that he was able to dissolve the crystals in a large amount of warm water.[60] The seemingly self-contradictory

59. For a review of the subject, see Malouin, "Sur le sel de la chaux," *HAS*, 1745 [publ. 1749], 93–106.

60. Gilles Boulduc, "Essai d'analyse en général des nouvelles eaux minérales de Passy," *HAS*, 1726 [publ. 1728], 306–327.

proposition that there could be salts insoluble in water was a new vista opened up by the alternative definition of the middle salt based on its mode of composition rather than on a traditional property of salts.

In a number of investigations carried out between 1734 and 1747, Duhamel pursued the question whether there is a salt contained in the alkaline earths. Because lime, unlike the other alkaline earths, was soluble in water but did not effervesce with acids, he thought that its active base might be different. He prepared a number of salts from lime combined with the mineral acids and with acid of vinegar. These salts, as well as salts of tartar, which he studied with another chemist of the Academy, Jean Grosse, turned out to have the same properties whether made with lime or one of the insoluble alkaline earths.[61] This work and similar experiments by other chemists led to a general acceptance by mid-century that the active base of all the alkaline earths was the same. It became known as calcareous earth.

Henry Guerlac has written that the impulse provided by French chemists in the Academy at the beginning of the eighteenth century, especially by Homberg and Etienne Geoffroy, soon faded out. The next generation of chemists in the Academy Guerlac dismissed as undistinguished.[62] It is true that there were no chemists of great stature in the Academy during the middle third of the century. Nevertheless, as the preceding examples suggest, lesser chemists pursuing problems framed by the classification of middle salts and the table of *rapports* maintained the Academy as a foyer for a powerful investigative thrust generated by Homberg and Geoffroy.

61. Duhamel and Grosse, "Des différentes maniéres de rendre le tartre soluble," *HAS*, 1732 [publ. 1735], 323–342; Duhamel, "Diverses expériences sur la chaux," ibid., 1747 [publ. 1752], 59–81.

62. Henry Guerlac, "Some French antecedents of the chemical revolution," in *Essays and papers in the history of modern science* (Baltimore, 1977), 340–341. Guerlac's view is another reflection of the influence of the canonical story discussed in the preceding lecture. The French chemists of this period made little contribution to those problems in chemistry most strategic to the forthcoming chemical revolution.

2. STAHL AND THE CHEMISTRY OF SALTS

I have not examined comprehensively whether chemists in other countries approached the chemistry of salts in the first half of the century as the French academicians did. The example of the most influential German chemist of the age, however, is suggestive of both strong correspondences and significant dissimilarities. Among his numerous writings, Georg Ernst Stahl published in 1723 a treatise on salts. His treatment of the same problem domain occupying the chemists of the Academy reveals a complex interplay of parallel, partially independent developments superimposed on a common core of shared procedures, empirical knowledge, and concepts; and an incomplete, somewhat erratic communication of influences back and forth between these two national specialist communities.[63]

The set of substances and operations that Stahl described in his treatise on salts coincided closely with those encompassed by Homberg's classification of salts and Geoffroy's table. Stahl too subdivided concrete salts into combinations of the mineral acids and acid of vinegar with fixed alkali, with volatile alkali, with earths, and with metals. He too described the orders in which acids, alkalis, and metals replaced one another in these combinations. The general displacement sequences—fixed alkali, volatile alkali, earths, and metals with an acid, and vitriolic acid, nitrous acid, and marine acid with an alkali—were identical to those in Geoffroy's corresponding columns. The order in which metals displace one another from an acid according to Stahl was sufficiently different from Geoffroy's order so that they must have been based on independent observations, but they were similar in principle. Stahl did not give a compact overview of these displacements equivalent to Geoffroy's table. He described the processes involved more fully, however, with more attention to exceptions to the general rules. For example, "although weaker than the other acids," marine acid "attacks silver, lead and mercury more preferably than nitrous and

63. Karl Hufbauer (ref. 20), 83–95, has argued that a self-conscious German national community of chemists did not emerge until the last third of the eighteenth century. The sense in which I use that term here does not require the degree of awareness and cohesion that Hufbauer has incorporated into his definition of the later German chemical community.

even vitriolic acid do."[64] In keeping with his long experience and great skill as a practicing chemist, Stahl discussed thoroughly the effects of varied conditions on the combinations and displacements and the properties of the salts formed, including their crystalline forms.

Like his French counterparts, Stahl viewed the substances formed by acids acting on alkalis, earths, or metals as combinations of these same constituents joined together. "Modern philosophy has reason to believe," he wrote, "that in compound bodies the parts composing them are inalterable in their nature and are only strongly attached to one another; and as long as the combination endures they have a different form and properties from those they each have separately."[65] Stahl's attitude toward these particular combinations, however, was quite different from that of the French chemists. Where they made middle salts central to the conceptual structure of the chemistry of salts, he regarded these combinations as the mere concrete material in which the component parts happened to be found. "When one talks of salts," he asserted, "one always has in mind their acid; but when one talks more grossly one has in mind the concrete form that they take in commerce; and then one includes the other parts that are contained in it, such as the metallic part in vitriol, the alkaline part in nitre, and the alkaline salt in marine salt." Those other parts, he stated in another place, serve merely to "trap" the "specific acid substance."[66]

As a practical chemist, Stahl was deeply familiar with the diverse properties of the several specific acids, including the differences in the nature and strength of their actions upon alkalis, metals, and earths. As a theorist, however, he wished to establish a common principle underlying the nature of acids in general. The primary acid he considered to be vitriolic acid, and he believed that he could prove experimentally "that this saline substance is formed by the combination of a very subtle earth and water."[67] The other acids

64. Georg-Ernst Stahl, "Traité des sels" (Paris, 1771), 230. This is an anonymous French translation of Stahl, *Ausführliche Betrachtung und zulänglicher Beweiss von den Saltzen, dass dieselbe aus einer zarten Erde, mit Wasser innig verbunden bestehen* (Halle, 1723). See J.R. Partington, *A history of chemistry, 2* (London, 1961), 662.

65. Stahl (ref. 64), 131.

66. Ibid, 44, 57.

67. Ibid, 48.

consisted in his view of vitriolic acid modified by the addition of varied amounts of phlogiston. Metals too Stahl sought to reduce to a common underlying combination of more basic principles: a mercurial part, an earthy part, and phlogiston.[68]

Stahl described acids sometimes as having "affinities" for metals. Since he also discussed the order in and relative force with which acids "attacked" metals and the order in which metals displaced one another in combination with acids, historians have sometimes assumed that Stahl was one of the founders of eighteenth century affinity theory. Stahl meant something very different, however, from what Geoffroy meant by *rapports*. Geoffroy was establishing empirical rules of combination and refrained from any explanation of why substances had stronger *rapports* with some substances than with others. Stahl used the word "affinity" as a synonym for "analogy," and intended it as an explanation for the fact that acids combined with or dissolved metals. The explanatory principle was that solvents attack those substances with which they share a common constituent, such as earth, the mercurial principle, or phlogiston.[69]

If the difference between Stahl's approach to the chemistry of salts and that of the French academicians appears on the surface to be conceptual rather than operational, it reflects an underlying difference in styles of practicing chemistry that can be traced in part to contrasting institutional patterns. Whereas the academicians reported the outcomes of their investigations to their colleagues in fulfillment of the obligation to advance their science, Stahl presented the results of his extensive laboratory experience mainly in a pedagogical context. Nurtured on the chemistry of the mid-seventeenth century, he began lecturing on the subject in 1684 at the newly founded University of Halle. Although he tried to put out a journal for about a year, he published almost all of his chemical observations and views in the form of long books that treated broad subject areas systematically.[70] His treatises included many original observations, but he did not organize them in the form of

68. Ibid, 293.

69. Ibid, 304–307.

70. Lester S. King, "Stahl, Georg Ernst," *Dictionary of scientific biography*, 12, 599–606; Partington (ref. 64), 659–662.

research reports. Moreover, when he discovered new processes or refined familiar ones, he presented his results less as contributions to an advancing science than as improvements to the craft. The chemists he appeared to be addressing in his writings were practitioners rather than investigators. He was much concerned about the superficial knowledge and poor operational standards of some of the practitioners.[71]

At another level, Stahl was engaged in the construction of an ambitious chemical system. Its best known feature was phlogiston; but it aimed more broadly at revealing the "intimate composition" of the tangible classes of substances—the acids, alkalis, metals, earths, sulfur, oils and others encountered in the chemical laboratory. This program was shaped by the writings of the chemists that Stahl had read in his youth, in particular Johann Becher. Even when writing in the 1720s, Stahl carried on a discourse largely with the great seventeenth century chemists Becher, Kunckel, and Glauber, referring only sparsely to an occasional more recent figure such as Homberg.

A successful university lecturer but a somewhat isolated figure, Stahl seemed, as he continued to argue with his early predecessors, out of step with contemporary chemists in the French Academy, who reacted to one another within a community of ongoing investigation. He had the advantage, however, of being a virtuoso laboratory chemist. His treatises were so rich in practical operations that he had devised to solve difficult problems that other chemists, including the French academicians, found in them many strategic leads for their own investigative ventures; or else found, as Duhamel did to his annoyance, after they had completed investigations they assumed to be original, that buried somewhere in Stahl's voluminous writings was an observation or conclusion that had anticipated their findings.[72]

Stahl was also a brilliant theorist. Although it was the phlogiston theory that caused French chemists to view him increasingly as the century progressed as the master architect of his field, his prestige also highlighted his broader program concerning the "intimate" composition of acids, metals, and other classes of substances. The

71. Stahl (ref. 64), 197, 224, 241, 267.
72. Duhamel (ref. 57), 230–232.

search for a small group of principles common to all substances, however, was an echo of an older chemistry becoming less and less relevant to investigative strategies based on the pragmatic concepts of middle salts and orders of displacements. The unusual pattern of Stahl's influence, peaking near the end of a long career and after his death, thus prolonged into the mid-eighteenth century conceptions of composition that were to some extent already anachronistic.

The conceptual structure within which Stahl worked and wrote was less conducive to a search for new acids, alkalis, metals, and earths, or their combinations, than to a search for the underlying composition of those already known. His treatise on salts dealt almost entirely with the repertoire of these substances passed on from the seventeenth century. The differences between his approach and that of the French chemists illuminates the significance of the research program emanating from the Academy. The classification of middle salts and the table of *rapports* were not mere generalized summaries of accumulated empirical knowledge about chemical substances and operations. They represented ways to view and to organize that knowledge and to set priorities for further investigation that differed strongly from the priorities embedded in Stahl's chemical system. Despite Stahl's authoritative position in eighteenth century chemistry, the program of the academicians proved to be more productive.

3. GROWTH OF THE CHEMISTRY OF SALTS

During the middle third of the century, German and French chemistry moved closer together, as influential French teachers such as Guillaume-Françoise Rouelle and Pierre-Joseph Macquer introduced Stahlian ideas into their courses and Macquer featured them in his popular textbooks; and as German chemists of the generations after Stahl integrated the pragmatic work of the French academicians with the systematic views of their mentors. Occasionally aspiring German chemists went to Paris to study with Louis Lemery, Simon Boulduc, or Jean Grosse. Most German chemists remained primarily practitioners and teachers in a growing number of university positions. In Prussia, however, the organization of the Berlin Academy in 1744 established salaried positions for two chemists to pursue full-time experimental activity, and provided them in 1753 with a new laboratory. Emulating the Parisian model, the Berlin Academy published annually a volume of memoirs. Its chemists were soon producing research papers similar in nature to the French academicians'.[73] The first two occupants of these positions, Johann Pott and Andreas Marggraf, proved to be resourceful research scientists who published frequently on the discoveries they had made while investigating central unsolved problems of the field. Marggraf was not only a rigorous analyst, but an imaginative investigative strategist, and he was soon recognized as the outstanding chemist in Europe.[74] Marggraf and Pott both made major contributions to the expanding field of the chemistry of salts.

Earlier I mentioned that by this period calcareous earth was identified as the single active base underlying the various common forms of alkaline earth. The procedures with which chemists identified the properties of calcareous earth also provided the means to recognize some alkaline earths that were distinct from it. For example, alum, one of the oldest known concrete salts, was generally regarded in the early eighteenth century as a compound of calcareous earth with vitriolic acid. In 1746 Pott pointed out, however, that any calcareous earth combined with vitriolic acid gave not alum but selenite (the barely soluble salt identified twenty years earlier by Gilles Boulduc). Taking up this problem in 1754, Marggraf tried to establish the composition of alum by separating its earth from its acid and then recombining them. Applying the procedure usually employed for separating an alkaline earth, he added fixed alkali to a solution of the salt and precipitated the earth of alum. Like calcareous earth, earth of alum dissolved in acids, but the salts it formed with them each had properties different from those that calcareous earth formed with the same acids. In reactions with other salts and alkalis, earth of alum showed additional distinct properties. Because the new alkaline earth could be obtained also from clay, it became known as argillaceous earth.[75]

73. Hufbauer (ref. 20), 45–48, 175, 181.

74. See Guerlac (ref. 62), 356–357.

75. Andreas Marggraf, "Expériences qui concernent la régeneration de l'alun," *HAS*, Berlin, 1754 [publ. 1756], 31–40; "Expériences faites sur la terre d'alun," ibid., 41–50; "Continuation des travaux sur la terre d'alun," ibid., 51–66; [Macquer] (ref. 35), *1*, 112–118.

Another alkaline earth with distinctive properties was magnesia alba, a white powder that Friedrich Hoffmann had precipitated from salt brine in 1722 by adding fixed alkali. With vitriolic acid it formed a bitter, soluble salt in place of the insoluble product of vitriolic acid and lime. In 1756 Joseph Black studied this earth further, and in 1759 Marggraf discovered it in several kinds of rock. Black and Marggraf both sought to prove that magnesia alba was a species of alkaline earth, but distinguishable in all its reactions from calcareous earth. Each based his case primarily on the observation that all of the salts he formed from magnesia alba with acids differed from the corresponding salts of calcareous earth.[76]

The emergence of chemically distinct alkaline earths, therefore, like the separation of the two fixed alkalis, involved the recognition and description of whole new groups of middle salts. Other new classes of salts became familiar through the discovery or characterization of metallic bases such as zinc, nickel, and cobalt. Chemists still relied mainly on the three common mineral acids and acid of vinegar to form these combinations, but a few new acids were beginning to appear. Phosphoric acid was identified as distinct from the common acids, and Marggraf formed salts of it with earth of alum and with magnesia alba to help demonstrate their specificity as new alkaline earths.

The characteristic acidic tastes of juices from citrus fruits, sorrel leaves, and other plant matters had long led chemists to believe that these juices contained acids distinct from the common "plant acid" obtained from vinegar. Sometimes these juices were employed as weak acids in various chemical operations; but they had not been isolated and were vaguely characterized. A similar situation regarding an acid contained in an oil expressed from ants prevailed until 1749, when Marggraf compared its reactions with various alkalis and minerals to the reactions of the common acids with them, and declared acid of ants to be a peculiar species resembling acid of

76. Partington (ref. 64), 696; Marggraf, "Démonstration fondée sur des expériences, que la pierre serpentine de saxe ne doit pas être mise dans la classe de l'argille," *HAS*, Berlin, 1759 [publ. 1766], 3–11; "Rapport des effets de l'acide du vitriol sur diverses pierres," ibid., 12–18; "Expériences chimiques concernant ce qu'on nomme la dernière lessive mère incristallisable du sel de cuisine," ibid., 19–27; "Expériences chymiques sur l'espèce de terre contenue dans la dernière lessive mère qui reste du sel commun," ibid., 1760 [publ. 1767], 75–86.

vinegar. Four years later Pott used similar means to demonstrate that the acid found in amber was different from any other known acid.[77]

The growth in the numbers of acids, alkalis, earths, and metals known was now beginning to impart to the chemistry of salts a self-reinforcing momentum. Each new identification of a member of one of the classes of substance increased the number of salts that one could attempt to form with any species of the categories that combined with that class. Each successive discovery consequently provided further research to do and a longer list of combinations to examine to make the identification of the new substance more certain. The growing complexity of the situation also demanded closer attention to all of the properties of each of the salts produced. Chemists made increasingly detailed observations of the crystals formed by a salt and of the conditions that affected their character. They still distinguished salts by older criteria, such as taste, whether or not they were deliquescent, and whether their acids could be separated by distillation. Increasingly, however, they characterized the salts by further changes that they underwent with other salts, acids, and bases. Marggraf reacted the new earths, for example, not only with the acids, but also with salts of the acids and other bases in order to extend his knowledge of their properties.

The relative solubilities of the salts also became steadily more important as a means to separate and identify them. Although the traditional idea that a salt is a substance soluble in water lingered on, the generalization of the meaning of the term salt to include the combinations of acids with all the types of substance with which they would unite induced chemists gradually to accept that the degree of solubility could range from very great to barely perceptible. Accumulating knowledge of the properties differentiating salts enhanced in turn the chances for discovering still more new acids, metals, alkalis, or earths contained in them.

The first edition of Macquer's *Dictionnaire de chymie*, published in 1766, offered a panoramic view of the chemistry of salts two thirds of the way through the eighteenth century. The entry "Salt" can be

77. Marggraf, "Observations sur l'huile qu'on peut exprimer des fourmis," ibid., 1749 [publ. 1756], 38–45; Pott, "Examen chymique de la nature de sel acide volatil de l'ambre," ibid., 1753 [publ. 1755], 51–72.

divided neatly into two portions, the first a critique of Stahl's "theory of salts," the other a survey of "the saline substances themselves and their combinations." Despite his great admiration for the views of the "illustrious" Stahl, Macquer weighed cautiously the question whether there was one universal salt underlying all saline substances; and whether, if that substance were vitriolic acid, it was composed of earth and water. "This grand idea," he wrote, "is by its generality capable of embracing and connecting together all of the phenomena that saline substances present; but we must at the same time concede that...it still falls far short of the facts and experiments necessary to give it the character of a demonstrated truth." Those "who desire the most sublime research in chemistry" he concluded this discussion, "cannot occupy themselves with a more interesting object."[78]

If, as Macquer seemed to imply, the experimental verification of the general theory of salts was a difficult challenge for the most "profound chemists," the chemistry of specific saline substances and their combinations was by contrast a field open to all enterprising investigators. There were, he wrote, "a very great number of known saline combinations," but still many more that had either not been prepared or had not been sufficiently examined. The combinations were now called "neutral" rather than "middle" salts. According to Macquer, neutral salts had earlier meant only substances "composed of acids and alkalis joined together to the point of saturation," so that neither the properties of the acids or alkalis were evident; but the term had now been enlarged to include "all combinations of any acid with all of the substances to which it can unite in such a way that it loses all or most of its acidity."[79]

Macquer listed the neutral salts in groups defined by the acid they contained. The first three groups, vitriolic salts, nitrous salts, and marine salts—that is, combinations of the three common mineral acids with alkalis, metals, and earths—each included seventeen specific salts. Tartarous, acetous, arsenous, and phosphorous salts formed smaller groups.[80]

78. [Macquer] (ref. 35), 2, 412–428.
79. Ibid., 2, 428, 430.
80. Ibid., 2, 430–441.

Macquer observed that some of the salts listed were "little known," "unknown," "little examined," or "only begun to be examined." He also mentioned whole groups, such as "plant salts"—that is neutral salts formed from acidic plant juices—"no member of which was known." He believed no doubt that these unknown combinations existed. Macquer's table of saline combinations was thus more than testimony to the remarkable growth of the chemistry of salts during the previous six decades; it displayed the heuristic role of this classificatory structure as a guide to future growth. The neutral salts listed as unknown were discoveries waiting to be made. Although he drew less attention to it, his classification also illustrated the fruitfulness of using neutral salts rather than acids as the central organizing framework. The acids of the tartarous and arsenous salts and salts of borax were not known in free form. Their existence was inferred from the existence of a series of salts composed of different alkalis, earths, and metals in combination with a common constituent. These categories of neutral salt invited research to try to isolate the assumed constitutive acids.

Drawing attention to the opportunities that beckoned, Macquer wrote enthusiastically:[81]

> One can see from this simple enumeration of saline combinations how many of them there are that are only very imperfectly known, how many that are not known at all, and even how many about which no one has ever thought. The numerous experiments that remain to be done in this vast part of chemistry are nevertheless a prime necessity, they are fundamental and elementary. To perform them successfully requires only accuracy, patience, and a knowledge of the first principles of chemistry. Every intelligent person of good will is capable of it. It requires one only to take well purified acids in order and to combine them with alkalis, earths, and metals equally purified; to examine the saline combinations resulting from these unions; to explore their taste, their solubility in water, their crystallization, their deliquescence, the way to decompose them, etc. These considerations are surely able to excite the zeal of those who desire to contribute their work to the progress of chemistry.

81. Ibid., 2, 440.

Believing that in this manner even beginners could multiply the discoveries of chemistry, Macquer appears to have underestimated the difficulty of such investigations. As the earlier example of Duhamel's experiments on fixed alkalis illustrates, to obtain a particular acid, alkali, earth, or metal in the required state of purity could test the analytical skills and the imagination of the best chemists of the age. Those challenges grew still more demanding as chemists took up such complicated situations as the acidic salts of tartar and the poorly characterized acids of plant matter. Nevertheless, the predictive power and experimental successes of the classificatory structure of the neutral salts continued to grow. The most stunning triumph of the eighteenth century chemistry of salts came during the next two decades, when the Swedish apothecary Carl Wilhelm Scheele isolated tartaric acid and then went on to isolate and characterize one after another the plant acids that until then had been only vaguely distinguished. More than doubling the total number of known acids, Scheele not only vastly enlarged the scope of the chemistry of salts, but established the first rigorously defined group of chemical substances within the domain of the future field of organic chemistry.

In their studies of the combinations of acids with alkalis, metals and earths, the chemists of the first two thirds of the eighteenth century not only achieved auspicious successes in their own time, but laid the conceptual and operational foundations on which much of the chemistry of the nineteenth century developed. By now it should be quite clear that these systematic chemical investigators were engaged in the construction of a modern scientific discipline.

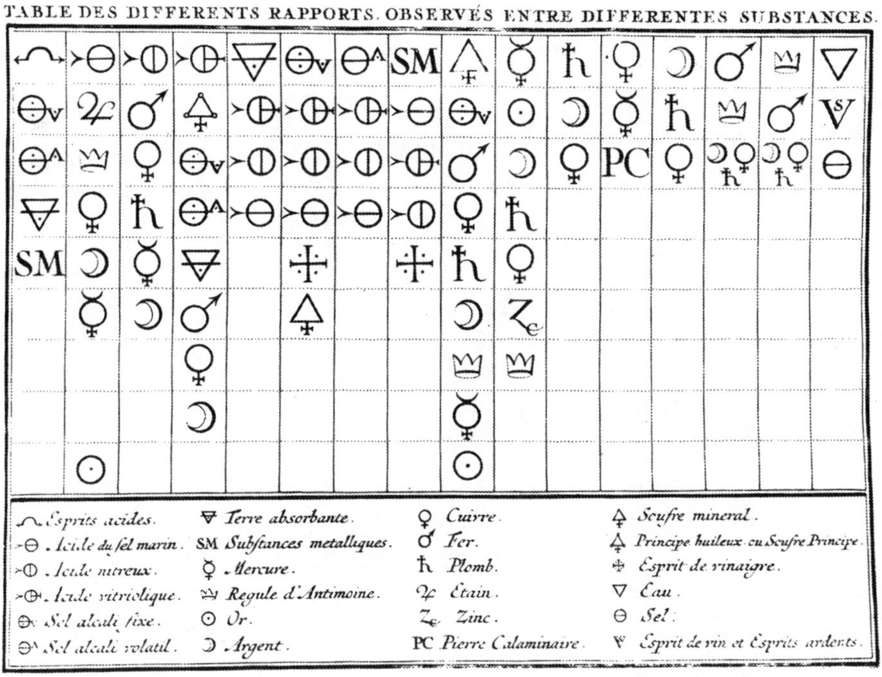

FIG. 12 "Table of the different *rapports* observed between different substances," from Geoffroy (ref. 49), facing p. 212. The first column gives the order of displacement of fixed and volatile alkali, absorbent earth, and metals as a group, in combination with any acid. Columns 2–4 give the order of displacement of specific metals in combination respectively with marine, nitrous, and vitriolic acid. The "oily" or "sulfur" principle, probably a pre-Stahlian version of an inflammable principle, appears next to the top in column 4. Column 5 gives the order of displacement of three mineral acids with absorbent earth. Columns 6 and 7 give the order of displacement of acids with fixed and volatile alkali, respectively. Column 8 gives the order of displacement of the acids with metallic substances. Column 9 gives the order of displacement of various substances in combination with sulfur. Columns 10–15 give the order of displacement of metals in combination with other metals. Column 16 gives the order of displacement of ardent spirits and salts in combination with water.

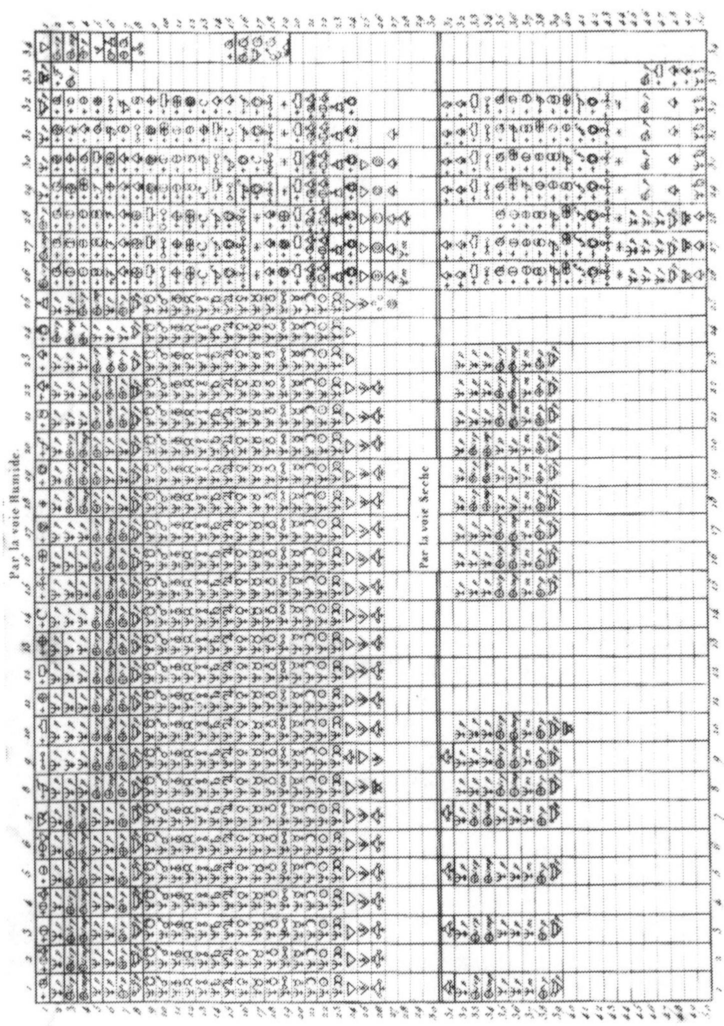

FIG. 13 Torbern Bergman's "Tables of elective attractions," from Torbern Bergman, *Dissertation on elective attractions [1775]*, trans. J.A. Schufle (London, 1968), facing p. xiv. The general format is simitar to that of Geoffroy's table. Comparison of the two tables gives a strong visual impression of the expansion of the chemistry of salts during the half century separating their publication.

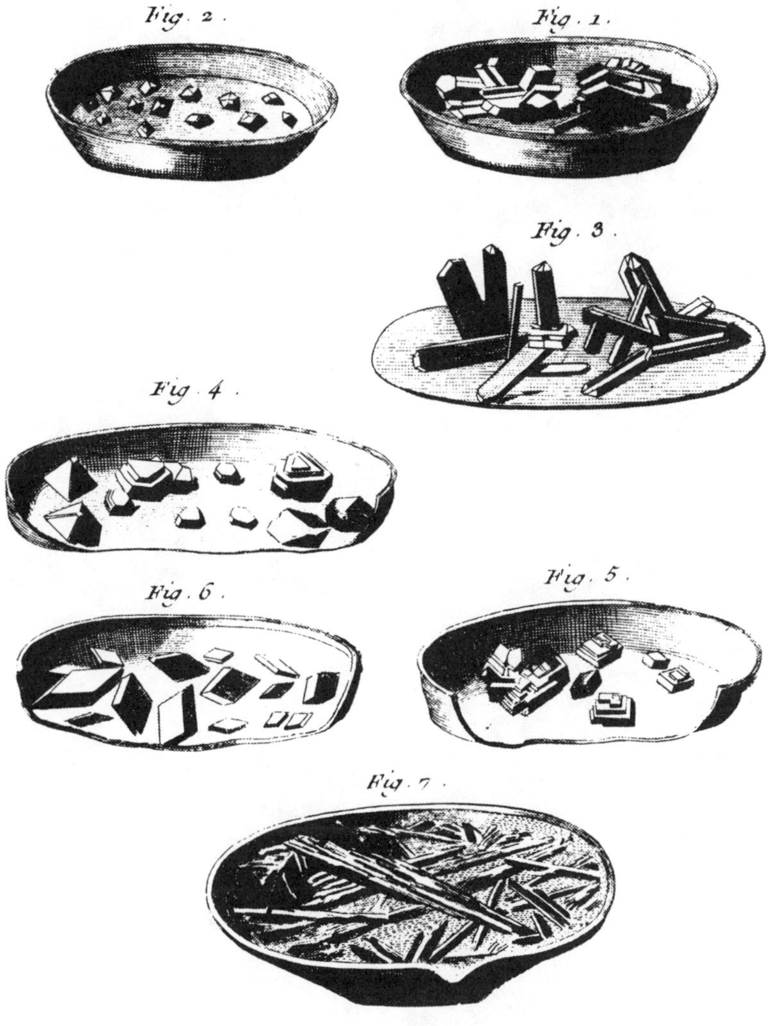

FIG. 14 "The crystallization of salts." *Fig. 1,* crystals of soda. *Fig. 2,* vitriolated tartar. *Fig. 3,* Glauber's salt. *Fig. 4,* alum. *Fig. 5,* green vitriol. *Fig. 6,* blue vitriol. *Fig. 7,* niter. Source as for fig. 1, plate xvii. The crystalline forms of salts provided a means to identify them and constituted one of the few effective methods available to eighteenth century chemists to obtain substances in relatively pure form.

3 CHEMISTS IN THE PLANT KINGDOM

Much of the practice of chemistry in the seventeenth century was ancillary to the preparation of drugs. The pedagogical framework within which chemistry emerged as a discipline during this century remained strongly attached to its medical objectives. Although Andreas Libavius was primarily a school teacher, the French teachers of chemistry who wrote popular textbooks descended from Libavius's great *Alchemia* were pharmacists who grounded their lectures and writings in the craft they practiced in their apothecary shops. Even as they sought to define chemistry as an independent art seeking knowledge of the composition of natural substances, their definitions tended to coalesce with the goals of identifying the medical virtues of these substances and of improving drugs through knowledge of their underlying principles.

Historians have assumed that one of the prerequisites for the formation of the modern science of chemistry was freedom from this subordination to medical purposes. Guerlac put the matter vividly when he described new directions that French chemistry took during the mid-eighteenth century as "the emergence of chemistry from its thralldom to pharmacy and medicine."[82] Among the purported consequences of such an emancipation was a shift of attention away from plant and animal substances too complex for eighteenth century chemists to manage to the simpler mineral substances whose composition they were capable of studying effectively. Subjecting the eighteenth century to the categories of a later era, historians have viewed organic chemistry as a subfield that could arise only after secure foundations had been established in inorganic chemistry.

82. Guerlac (ref. 62), 103.

Like much else that I have discussed in the first lecture, these judgments are oriented around the chemical revolution as the culmination of all eighteenth century chemistry. The fact that Lavoisier was neither medically trained nor involved with pharmacy; that he entered chemistry through geology and mineralogy; that he is perceived as having achieved his revolution by concentrating on the combustion of simple mineral substances; and that he wished to make chemistry more like experimental physics, all reinforce a historical bias toward the older association of chemistry with medicine as an obstacle that had to be overcome.

If we view eighteenth century chemistry instead as a pluralistic problem field in which development took place in several domains, not all of which were at issue in the chemical revolution, then we can reach a more positive perspective on the relation between chemistry and medicine. Most of the chemical teachers and investigators of the eighteenth century were also physicians or pharmacists; and during the same period in which mineral chemistry was advancing along the lines outlined in my last lecture, these pharmacist-chemists were transforming the modes of investigating plant and animal matters.

Eighteenth century chemistry was divided not into the two subfields of organic and inorganic chemistry, but into three parts based on the traditional three kingdoms of nature. In keeping with the practice of identifying substances according to their sources rather than their compositions, seventeenth century textbooks grouped chemical procedures into operations carried out on plant matter, animal matter, and minerals. This custom was retained long after it became evident that some of the same substances could be obtained from two, or even all three of the kingdoms. Within this schema, the chemistry of the plant kingdom maintained a pre-eminence grounded in the rich variety of medical "virtues" procurable from plant matters—especially from exotic plants—and in the experience that the traditional distillation methods appeared to separate more readily from plant matters than from those of the other two kingdoms the water, spirits, oils, salts, and earth thought in the seventeenth century to comprise the elementary principles of all matter.

1. PLANT ANALYSIS AT THE PARIS ACADEMY OF SCIENCES

Among the investigative challenges facing the chemists mentioned in the last lecture who entered the French *Académie des Sciences* between 1690 and 1706, the most formidable was to decide what to do about a large set of laboratory notebooks recording analyses performed by their predecessors on hundreds of plant materials. In 1670 the founding members of the Academy had undertaken an extraordinarily ambitious project to write a natural history of plants. Pooling their talents, the Academicians intended to describe in great detail the anatomy of all the plants they could gather from all over the world, to investigate their "vegetation" by growing them, and to discover their unknown medical virtues, if necessary by testing them on patients. The chemists were to establish the relation between the virtues of each plant and its composition.[83]

Relying on a basic mode of analysis that herbalist distillers had employed for over two centuries to obtain drugs from plants, one of the Academy's chemists, Claude Bourdelin, subjected each plant or its several parts to a prolonged distillation in which he collected in receiving flasks, changed at frequent intervals while he gradually raised the heat, a series of acid spirits, oils, volatile alkali, and water. Afterward he examined the fixed salts and the earth that remained in the distillation vessel. Well aware of the pitfalls attending such analyses, the academicians went to great lengths to impose rigorous scientific standards on this traditional method. Knowing that the distillation of all plant matters yielded qualitatively similar products, they sought to differentiate them by measuring accurately the quantities of each type of product. Knowing that variations in the conditions of a distillation caused differences in the qualities and quantities of the products received from two analyses of the same plant matter that were often greater than the differences in the products from dissimilar plants, they developed a standardized procedure that they applied uniformly to every plant material they examined. Sensitive to an objection often raised against such analyses, that the fire used in distillations did not

83. Denis Dodart, "Mémoires pour servir a l'histoire des plantes," *MAS*, 1666–1699, *4* (Paris, 1731), 121–242, esp 121–151.

merely separate the parts constituting the plant matter, but rather modified or even destroyed the original constituents, and that the products received might be creations of the fire, the academicians sought to minimize such effects by employing at each stage of the distillation the least degree of heat necessary to carry out the operation. To identify acids and alkalis, they applied the indicator tests devised by Robert Boyle. To estimate the "strength" of the acids and alkalis obtained, they studied the effects of acids and alkalis of different strength on the colors of the indicators.[84]

Armed with such methods, Bourdelin analyzed in the laboratory of the Academy dozens of materials each year. Sometimes he fixed his attention on special problems. In 1675 he found a way to obtain oils in purer form. In 1680 he analyzed animal matters such as blood and lymph. As he performed year after year standardized distillations on all the plants that came to the Academy, Bourdelin settled into a routine that went on and on of its own momentum, ending only with his death in 1699.[85]

Soon after Wilhelm Homberg was named a member of the Academy in November 1691, its recently appointed director, the Abbé Bignon, ordered him to consider how to make use of all of the work that the Academy had done on the analysis of plants. Denis Dodart, who had played a leading part two decades before in defining the project, turned over to Homberg three volumes of the laboratory record of the analyses. In September 1692, Homberg read a report of his conclusions to the weekly assembly of the academicians. The work, he said, had been prodigious both in quantity and in accuracy. He found "a great uniformity" in the substances composing all of the plants analyzed in these volumes, so that their differences seemed to him to "consist for the most part only in the different combinations of these materials." Although the same materials were found in them, there was "always a difference in the quantities of one or the other" of the constituents.

84. Ibid., 152–242.

85. *HAS*, 1666–1699, *1* (Paris, 1733), 18–20, 56–58, 120–122, 163, 196, 203–207, 209–221, 252–253, 282, 307–308, 328–329, 373, 405–406; *2*, 9–10, 26–27, 49–50, 68, 122, 182–183; "Éloge de Monsieur Bourdelin," *HAS*, 1799, 2nd ed. (Paris, 1718), 122–123; [Bourdelin] "Reçueil d'analyse chymique," unpublished laboratory notebooks, 11 vols., Bibliotheque Nationale, Paris; Procès verbaux de l'Académie Royale des Sciences, vols. 4–18, Archives of the Académie des Sciences, Paris.

"That is not to say," he acknowledged, "that one does not sometimes encounter plants that greatly resemble one another in the analysis that nevertheless have opposite effects when used in medicine." Consequently, "one would not be able to judge the effect of a plant entirely by its analysis."[86]

Homberg had found in the laboratory notebooks "experiments that establish truths formerly contested or even denied;...for example, the volatile salts of many plants and the different lixivial salts" they contained. These were, he averred, "very considerable discoveries that should please savants; but the public should also find in this work some particular utility in medicine."[87]

Some historians have used portions of Homberg's statements out of context to support the view that the herculean efforts of the earlier academicians had been "tedious" and "useless."[88] That was clearly not Homberg's opinion. He believed the work had not only been impressive scientifically, but had included significant discoveries. What remained in doubt was its usefulness for medicine. His view reflected not only the continuing close association between experimental chemistry and medical practice, but also the pressure that the French state maintained on the academicians to show that their scientific work had social applications.

In order to discover how the results attained by his predecessors could be made more useful, Homberg fixed his attention on the "fetid oils that were found at the end of each analysis." By distilling such oils repeatedly with quicklime, he was able to remove the "burnt" parts responsible for their disagreeable qualities, leaving an "extremely penetrating oil" that he employed to good effect in the treatment of paralytic and rheumatic pain.[89]

During the following years, Homberg continued to investigate problems related to the analysis of plants, moving from his initial concern with the medical utility of the results toward more basic questions about the composition of plants. In 1695 he pointed out that of all the substances to which plants are reducible by

86. Ibid., "Procès verbaux," *13*, 69v, 115v–116v.

87. Ibid., 116.

88. J.R. Partington (ref. 64), *3*, 12; Boas (ref. 37), 73; A. Rupert Hall, *From Galileo to Newton, 1630–1720* (New York, 1963), 147.

89. Procès verbaux, *13*, 116v–117v; *HAS*, 1666–1699, *2*, 148–150, gives a slightly abbreviated version of Homberg's report.

distillation, only the earthy residue seemed to be of the same nature for all types of plant matter. The great differences in the taste, odor, and consistency of oils suggested to him that they might be further resolvable into simpler constituents. Having found in his earlier distillations of fetid oils with quicklime that an odoriferous water appeared, he now concluded that the water must have been one of the constituents of the oil. Extending the process to greater heats, he was able to convert nearly all of the oil to water. The remaining residue turned gray because, he thought, it retained the salts and earth of the former oil. From this and other accessory evidence, he concluded that oils are composed of water, salt, and earth.[90]

In 1700 Homberg presented to the Academy the outcome of a resourceful methodological investigation in which he verified his prediction that a strong mineral acid added to plant matter enabled one to distill from the plant considerably more oil than a standard distillation yielded.[91] The next year he broadened his approach to confront the most fundamental theoretical issue surrounding the distillation method of plant analysis: "Whether the...materials into which a mixt [the contemporary term for a substance composed of simpler substances] is reduced by its fire actually exist in the mixt when it is in its natural state." He mentioned two reasons why some people had come to doubt that they did. The first was that "two plants completely different in flavor, odor, form, and virtues, such as for example *Solanum furiosum* and the *Brassica capitata*, are reduced by the analysis into principles so similar in number and quality that one would accept them as [the results of] a single plant analyzed twice, although one of them is a poison, and the other is a kitchen vegetable."[92] This more colorful version of the idea he had already expressed nine years earlier was taken by some later commentators to mean that the analyses proved nothing at all. Homberg used this extreme example, however, only as an objection raised by others that he had to answer. The other objection against the products being the true principles that he acknowledged was

90. W. Homberg, "Sur les huiles des plantes," *HAS*, 1666–1699, 2, 246–249.

91. Homberg, "Observations sur les huiles des plantes," *HAS*, 1700, 2nd ed. [Paris, 1719], 212–217.

92. Homberg, "Observations sur les analyses des plantes," *HAS*, 1701, 2nd ed. [Paris, 1719], 115–119, on 115.

that no one had been able to recompose the original mixts by combining the principles separated from them.

Homberg sought a middle ground, similar to what his predecessors had defended, between the polar opinions that the fire merely separated and that it created the products of the analysis. He provided, however, a fresh experimental justification for this stand:[93]

> I say that we cannot deny absolutely that the four materials, that is, salt, water, oil, and earth, enter into the composition of a plant, since one always finds them no matter in what manner one makes the analysis; but that the doubt consists solely over knowing whether they are in the same form in the plants as the great fire gives them to us in their analysis, or if the fire alters the principles and if it causes us to see them differently from what they are in the mixt.

Homberg believed that by comparing the results of varied analytical procedures he could clarify how the principles contained in plant matter may be altered in their separations. Of the experiments that he performed with this objective in mind, he described only one example in his published memoir. Juice expressed from grapes and distilled fresh in the usual way gave off first an insipid phlegm, then several acidic portions, some volatile salt, and a foul oil, leaving a lixivial salt and earth. If Homberg instead evaporated the same juice over a small fire to one third its initial volume, a slightly acid essential salt precipitated, and sweet, pleasant-tasting oil appeared on the surface. Allowing some of the same juice to ferment instead until it turned to wine, he distilled spirit of wine and water from it. Then he evaporated the remainder to the consistency of honey and mixed it with some of the spirit of wine. The latter extracted from it a red aromatic oil, while an acid salt resembling tartar crystallized and a little earth precipitated.[94]

These three different analytical methods had, in Homberg's view, yielded the same basic principles, except for the fact that in the second process the earth he assumed the juice to contain could not be separated from the oil and the salt. Since this process had required neither a large amount of heat nor fermentation, he considered the principles least altered in the second case, an inference

93. Ibid., 116.
94. Ibid., 116–117.

he confirmed from their tastes; the oily matter retained the sweetness of the fruit, and the salt still resembled the original state more closely than did the products of the standard distillation. Using the second process as a reference, therefore, he could deduce something about the nature of the changes wrought by the other two. He believed that the natural salt was a complex one that the heat decomposed into an acid, a volatile salt, and an earth. He guessed that the sweet oil became through the heat volatile and smelly by being joined to some of the acid and the volatile salt. The best way to discover the "true principles and virtues of a plant," Homberg concluded, was to avoid strong fires, employing gentle heat together with fermentation, which "aids the natural separation of the principles that compose these" materials.[95]

A decade of work on plant analysis thus led Homberg to the conservative conclusion that the customary distillation methods, supplemented in special cases by other procedures, could lead to a knowledge of the composition of plants. Generations of pharmacists and chemists before him had affirmed the importance of avoiding too much heat. Homberg also maintained a traditional view of what the principles composing any mixt body were, eliminating only one from the conventional list of five. He was, however, not merely validating the past. Open-minded and experimentally flexible, he was raising more questions than he resolved, endeavoring through new combinations of well-known approaches to advance the state of the problem.

2. THE ANALYTIC PROGRAM OF SIMON BOULDUC

At the first meeting of the Academy in the eighteenth century—January 9, 1700—Simon Boulduc reported on analyses he had made of ipecacuanha, a root that had been found to have a specific action against dysentery, as well as general emetic, cathartic, and astringent effects. After describing the three types of ipecacuanha, he wrote:

> I began by working on the gray [variety], resolving to continue my

95. Ibid., 117–119.

> work on the other two. In order to commence my analyses I first took the ordinary way, which is that of distillation, proposing to myself, however, to go on from there to other [methods] in order to complete my observations and my research. Therefore, following our regular order, and degrees of fire, I operated on eight ounces of that root, and I collected three portions.

Boulduc then described the liquids collected successively in the three receivers, the dry black matter left in the retort, and the fixed fermentable salt he had extracted from this residue after calcining it. He also had applied "our ordinary tests" on each of the substances obtained, but thought the results too inconsequential to detail.[96]

From this account it appears that Boulduc had undertaken his investigation by following the standardized operative procedures that Bourdelin had employed for the analysis of plants for nearly thirty years. Since Bourdelin had died just four months earlier, it is plausible that the Academy had even expected Boulduc to carry forward its long-standing project, although Boulduc had already planned the previous spring a large work on purgatives as his personal project. At any rate, he quickly became disenchanted with his initial approach:[97]

> But since by this distillation analysis ordinarily employed to decompose mixts, I found nothing but the proportions of the five principles, and my mixt was nevertheless destroyed without enabling me to recognize its virtue either in its separated parts or in the same parts recombined...I therefore believed that I must proceed by another way, or by another kind of analysis, which I would call analysis by extraction, through which I could succeed in drawing from this root some essential part in which I could locate the specific virtue such as it appears to us in the intact mixt.
>
> Taking this view, and being very persuaded, moreover, that the force of this root can only consist of its resinous part...or its saline part, I believed that I needed only to search for analogous and convenient solvents in order to extract each of these parts and to perform afterward experiments in order to ascertain in which of them its virtue, whether emetic or purgative, mainly resides. For that purpose I prepared several infusions and digestions of eight ounces of gray ipecacuanha with very rectified spirit of wine, the ordinary solvent for resinous parts.

96. Procès verbaux (ref. 85), *19*, 1–2.
97. Ibid., *18*, 143v, 510–510v, and *19*, 22v.

After dissolving out the resinous portions in this way, Boulduc infused the remaining residue in distilled water to extract the "saline" substances. Reversing the order, he extracted first with water, and found that afterward spirit of wine extracted very little more material, from which he inferred that gray ipecacuanha contains much more saline than resinous matter. Boulduc tried out the extracts obtained with spirit of wine and water on a number of patients suffering from unspecified maladies, and found in some cases that the patient recovered, or at least that the extract promoted the evacuations so critical in prevailing medical practice.[98]

On the immediate face of it, Boulduc had done nothing very new. Spirit of wine and water had long been used to extract resins and salts respectively from plant materials. Nevertheless, his investigative narrative, contained in the transcript of Boulduc's presentation to his fellow academicians (and presented in compressed, less revealing form in his published memoir on the subject) captures the nascent stage of a development that we can see retrospectively as a watershed in the history of plant chemistry.[99] What was most significant about Boulduc's move was that he had broken away from the deeply entrenched standard order of analysis, putting in its place a general procedure of analysis by extraction employing methods that had formerly been applied only in special cases. Having taken this turn, he pursued his new investigative course with dogged persistence.

Continuing his experiments on ipecacuanha, Boulduc was able to distinguish the composition of the gray and brown varieties in terms of the proportions of the extracts they yielded to the solid residue they left. In 1701 he applied his extractive methods to four more purgatives: coloquinth, jalop, *gomme goutte*, and the celebrated powerful but dangerous black hellebore. In 1705 he examined the native French plant gratiola, and in 1708 the juice of aloes. The next year he took up cachou, a thick juice from Asia, followed in 1710 by rhubarb and in 1714 by agaric, a mushroom. Yielding to the weight of custom, he routinely repeated the ordinary distillation analysis on each plant, even though he regularly reported that it

98. Ibid., *19*, 2v–6.

99. Boulduc, "Analyse de l'ypecacuanha," *HAS*, 1700, 2nd ed. [Paris, 1719], 1–5, on 3–4.

had revealed nothing instructive. With his extractions, on the other hand, he often varied his basic procedure in order to increase the amount of substance drawn out or to explore the resulting variations in its properties. Some of the gums that would not dissolve in spirit of wine he found he could extract in a solution of salt of tartar or other alkali. Distilled vinegar proved a very effective solvent for some substances. He experimented extensively to ascertain the relative effectiveness of water at different degrees of heat.[100]

Properties such as color and taste, by which the products of analysis had traditionally been judged, often served Boulduc as clues to whether or not his products embodied the medical virtues he sought; but he regarded such criteria as misleading and regularly tested their properties by administering them "with all the necessary precautions and discretion" to patients.[101] From such experiments, he reached the general conclusion that the purely resinous substances extracted with spirit of wine purged more violently than the crude materials; that the purely saline extracts had little purging action, but were diuretic; and that saline and resinous portions mixed together purged most gently and usefully. On the basis of this experience, he could attempt to improve drugs rationally, retaining those actions he wished to have while eliminating undesirable effects.

In Boulduc's view, the principles composing a mixt should be recognizable by their medical effects. Substances obtained by extraction retained properties characteristic of the original material, whereas those derived by distillation possessed no special actions distinguishing the products of one mixt from those of another. His results proved, he asserted, that the products of distillation, "which are improperly named principles, retain none of the virtues of the mixt from which they are drawn: it is not the same for those that

100. Boulduc, "Suite des analyses de l'ypecacuanha," *HAS*, 1700, 2nd ed. [Paris, 1719], 76–78; "Observations analitiques du jalap," ibid., 1701, 108–111; "Observations analitiques de la coloquinthe," ibid., 1701, 12–17; "Remarques sur la nature de la gomme gutte et ses differentes analyses," ibid., 1701, 133–137; "Observations sur la gratiole," ibid., 1705 [publ. 1706], 186–194; "Sur l'aloes," ibid., 1708 [publ. 1709], 54–55; "Observations et analyses de cachou," ibid., 1709 [publ. 1711], 227–232; "Observations sur la rhubarbe," ibid., 1710 [publ. 1712], 163–169; "Sur l'agaric," ibid., 1714 [publ. 1717], 27–30.

101. Boulduc, "Analises de la coloquinte, du jalap, de la gomme gutte," *HAS*, 1701, 2nd ed. [Paris, 1719], 58–62, on 59.

various extractions give us: we know that the products that result from them comprise, as in an epitome, all of the active principles of the mixt."[102]

Wilhelm Homberg and Simon Boulduc must have performed some of their analyses on plant matters at the same time in the same laboratory of the Academy of Sciences. The juxtaposition of their contrasting investigative trajectories is pertinent to the general themes I outlined at the beginning of this lecture. The starting point for both was the massive project of plant analysis carried on by their predecessors, but their relations to it were different. Homberg was asked to evaluate what had already been done. Boulduc tried to follow the same operating procedures employed in the project for his own special research subject, purgative drugs. At the beginning, Homberg, like Boulduc, was concerned with the medical usefulness of such analyses; but whereas for Boulduc the concern was intrinsic to his own central purpose, to Homberg it was probably a duty imposed by institutional circumstances. As Homberg pursued his problem, he immersed himself increasingly in more general questions about the composition of plant matters and the adequacy of the distillation analyses to uncover that composition. Boulduc maintained all the way through his research the priority of medical criteria of composition.

Homberg has been singled out by historians as the most prominent chemist in the Academy in this period. He is perceived as a broadly trained, versatile scientist. He, if anyone in that setting, would be expected to contribute toward the emergence of chemistry as an independent, progressive discipline. Boulduc, on the other hand, little known to historians, an apothecary to royalty, conservative in his broad orientation, might well appear to symbolize in the Academy the supposed older subservience of chemistry to medicine. In the realm of plant chemistry, however, it was Homberg who held more closely to the traditional order of analysis and Boulduc who opened up a new approach to the problem. He did so because, rather than in spite of, his medical orientation. As Homberg's assessment of the project of plant analysis shows, he recognized but was not personally limited by the inability of chemical analysis to differentiate plants according to their medical

102. Boulduc (ref. 99), 5.

properties. Within the bounds of current analytical methods, he saw room to pursue more basic chemical questions. For Boulduc, however, the failure of the standard methods to reveal the principles within which the medical virtues resided was critical. It was that failure that drove him toward the most innovative investigative pathway followed by a member of the talented group representing chemistry in the Academy of Sciences at the beginning of the eighteenth century.

3. THE METAMORPHOSIS OF A TRADITION

The sharp contrast that Boulduc drew between the positive results of his new analytical program and the unilluminating results of his standard distillation methods did not immediately induce other chemists, even within the Academy, to give up on the older approach. The elder Lemery continued unperturbed to publish investigations of honey and urine using the standard procedures.[103] For the younger chemists, Homberg's comment that the many analyses carried out by Bourdelin had not revealed any difference between the substances obtained from an edible plant and a poisonous one rendered the distillation analyses problematic; but instead of abandoning the conventional method they sought, as Homberg did, to overcome its limitations. Freshly admitted to the Academy, Claude-Joseph Geoffroy hoped in 1707 to be able to distinguish the distillation products from different plants by devising new chemical tests capable of differentiating them. Mixing oil of thyme with each of the common acids and alkalis, he was able to produce a series of dramatic color changes. Unfortunately, none of the other essential oils he tested showed similar effects, and he was unable to draw any general conclusions.[104]

In 1719 Louis Lemery criticized the ordinary method of analysis more harshly than had either Homberg or Boulduc. The fire,

103. N. Lemery, "Du miel et de son analyse chimique," *HAS*, 1706 [publ. 1707], 272–283; "De l'urine de vache, de ses effets en médecine, et de son analyse chymique," ibid., 1707 [publ. 1708], 33–40.

104. C.J. Geoffroy, "Observations sur les huiles essentielles, avec quelques conjectures sur la cause des couleurs des feüilles et des fleurs des plantes," *HAS*, 1707 [publ. 1708], 517–526.

Lemery stated, "does not spare any of the substances submitted to its action, but...crushes them," giving rise often to "new compounds that differ very drastically from those that inhere naturally in the mixt."[105] Consequently the old project had failed to reach its objective. Instead of repudiating the method itself, Lemery drew implicitly on the conception of middle salts (or "concrete" salts as he called them) that had emerged in the Academy since the time of the project. He interpreted the chemical changes produced by the distillations as separations of acids or alkalis from middle salts that had existed in the plant matter, or new combinations formed by the mixing of acids and alkalis separated in the original materials. He concluded optimistically that it would be possible to infer in this way from the products and conditions of the analyses the natural state of the acids, bases, and salts in the plant.[106]

The most influential teacher of chemistry in the early eighteenth century was Hermann Boerhaave, whose lectures at Leyden attracted students from all over Europe. In 1732 Boerhaave wrote a large textbook summarizing what he had been teaching for thirty years. The second of its two volumes was devoted largely to plant chemistry, and in it he described analytical processes that were by then traditional. He believed, along with his predecessors for more than a century, that the agent that came closest to the ideal of separating the parts of a plant "body" without altering them was "a gentle action of fire." Well aware that in practice the distillation procedures employing fire were incapable of removing all the parts of plants without changing them, or of separating all of the medical virtues they contained, Boerhaave utilized distillations pragmatically. When he could attain by distillation an oil retaining the characteristic taste or scent of a plant, or an appropriate medical virtue, he did so. In other cases he resorted to processes such as decoctions. The fact that chemical analyses could not preserve all of the "pure and perfect virtues" of plants, but that some mixing

105. L. Lemery, "Reflexions physiques sur le défaut et le peu d'utilité des analises ordinaires des plantes et des animaux," *HAS*, 1719 [publ. 1721], 173–188, on 174.

106. Lemery, "Second mémoire sur les analises ordinaires de chimie," *HAS*, 1720 [publ. 1722], 98–107; "Troisième mémoire sur les analises de chimie, et particulierement sur celles des végétaux, ibid., 166–178; "Quatrième mémoire sur les analises ordinaires des plantes et des animaux," ibid., 1721 [publ. 1723], 22–44.

and changing always occurred, did not deter him. Cautious and precise, he was content to acknowledge the "narrow bounds" of the art of chemistry, while refining the conventional procedures to make them as rigorous as possible. His most striking refinement was to employ the new Fahrenheit thermometer to attain more consistent results by controlling more accurately the degree of heat during a given operation.[107]

A mixture of conservatism with a sensible belief that the potential for improving the customary distillation methods for plant analysis was not yet exhausted may explain the slowness of early eighteenth century chemists to adopt Boulduc's approach. Over the course of the first half of the century, however, we can trace a gradual shift in attitudes and practices. Distillations remained prominent, but the old "order of analysis" centered around a graduated distillation receded in importance. Meanwhile extractions with spirit of wine and water were more broadly employed. It is difficult to assess the degree of influence that Boulduc himself had on this trend. I have so far found no explicit references to his work in this connection in the subsequent contemporary literature. Yet his work must have been known within the intimate circle of chemists in the Academy, and he was prominent enough that German chemists included him among those with whom they came to study in Paris. Reinhard Löw has suggested recently that his methods may in this way have had an impact on one of the leading German students of Stahl, Caspar Neumann.[108]

Sent abroad twice to study chemistry, Neumann traveled in 1719 from London to Paris, where he attended chemical lectures, taught a course himself, and carried out experiments twice weekly with the Geoffroy brothers. It would be improbable that he would not also during this extended stay have had contacts with Boulduc. In Berlin during the next 17 years as court apothecary, Neumann published, among other things, an analysis of camphor in 1727, one on opium in 1730, and one on coffee and tea in 1735. Löw has emphasized that Neumann relied during these investigations on

107. Hermann Boerhaave, *A new method of chemistry*, trans. Peter Shaw, 2nd ed., 2 (London, 1741), 3–170.

108. Reinhard Löw, *Pflanzenchemie zwischen Lavoisier und Liebig* (Straubing, 1971), 41.

double extraction methods with water and spirit of wine to separate the constituent substances from these plant materials, and that he was particularly progressive in subjecting the products so obtained from coffee and tea to a series of reagents—acids, alkalis, and salts—and noting the resultant color changes or precipitations. Neumann, like Boulduc, dismissed distillation analyses of these substances as unable to provide information useful to his purpose.[109]

In his general chemical lectures, on which several published works were based, Neumann enumerated "the substances naturally contained in vegetables, and separable from them by art" as "(1) gum, (2) resin, (3) a subtle oil, (4) camphor, (5) a gross oil, and (6) a saline matter, together with phlegm or water...and (8) an earthy substance that is the basis of all vegetable bodies." Neumann defined these substances through a list of properties that featured their solubilities:

- resins are soluble in vinous spirits and oils, not acted upon by water.
- subtle oils are sapid and odorous, soluble in spirit of wine, miscible with the grosser oils, not miscible with water, totally volatile in the heat of boiling water, very readily inflammable.
- camphors are solid concretes, soluble in vinous spirits, in oils, and in mineral acids...not acted upon by water, totally volatile in the heat of boiling water, easily fusible, highly inflammable, subliming unchanged in closed vessels.
- the gross oils are insipid and inodorous, not miscible with water or with vinous spirits, but readily with essential oils, not volatile in the heat of boiling water, inflammable in the fire.

Saline bodies and earth were also defined, as they had always been, in part by their solubility and insolubility, respectively, in water.[110]

Neumann applied methods of separation appropriate to the salient properties of each of these types of substance. Gums and resins he obtained mainly through extractions in water, spirit of wine, or both used in succession. To procure essential oils and camphor he distilled plant materials in water, since both substances were volatile enough to be received without losing their

109. Ibid., 57–60; Karl Hufbauer, "Newmann, Caspar," *Dictionary of scientific biography, 10,* 25–26.

110. Caspar Neumann, *The chemical works,* abridged ed. William Lewis (London, 1759), 264–265.

characteristic properties. Essential salts, "obtained chiefly from plants of the acid sort," he got by expressing the juices of the plant, and allowing them to stand until a deposit separated. Then he filtered the latter out and evaporated the filtrate until crystals formed. He washed the crystals and recrystallized them until they were pure and white.[111]

Throughout his procedures, Neumann paid careful attention to the conditions required, whether in extractions or distillations, to preserve unaltered the qualities of the *native* principles, by which he meant those principles that "naturally exist in the subject, art doing nothing more than barely to extract them." These principles were clearly not "simple bodies for they are all compounded;" but, "as it is by these that one plant differs from another, and that every plant exerts its specific powers..., we cannot further resolve them without totally destroying their peculiar properties."[112]

Briefly Neumann listed the traditional products obtained from plants "distilled in close vessels by a strong fire." These products, he declared,[113]

> have no resemblance in taste, smell, or any other sensible property, to the original subject nor are [those]...produced from one vegetable to be distinguished from those afforded by another. Yet such has been the infatuation of the chemists that in examining particular plants they have hitherto fondly pursued this destructive resolution, and assigned, for the natural principles of the plant, these artificial products that had no existence in it.

As these passages imply, Neumann distanced himself from the traditional distillation procedures for motives similar to Boulduc's. They did not preserve specific properties characteristic of particular plants. The criterion was a practical rather than a theoretical one, closely related to the medical or other uses for the substances separated. Neumann merged these needs, however, with the general question of the composition of plant matter by defining the substances retaining these properties as the "native principles" of the plant. This was a significant revision of the earlier meaning of "principles," which had customarily been applied to the simplest

111. Ibid., 267–280.
112. Ibid., 280.
113. Ibid., 280–281.

parts into which a substance could be resolved. Traditionally it had been permissible to think of these principles as both simple and variable from one source to another. The term "principle" was traceable back to the Paracelsian three principles, which were conceived both as elementary and protean in their manifestations. Strong residues of this ambivalence remained in the eighteenth century, but the growing rigor with which specific substances such as acids, bases, and salts could be identified and distinguished from others of the same class made it increasingly difficult to ascribe both variability and an elementary status to the same substance. Neumann clearly implied that "native principles" could retain properties characteristic of the many specific plants from which they were derived only because the principles were *not* simple.

During the second quarter of the century, French chemists too moved away from the efforts of the first decades to improve the customary distillation procedure, and came to view that general order of analyses as part of a past age. This movement took place on two levels—the investigative forefront and the application of the results of such analyses for medical practice.

The only chemist in the Academy who pursued experiments in plant chemistry persistently through this period was the younger Geoffroy, Claude-Joseph. Between 1721 and 1730, Geoffroy carried out a narrowly focused but probing series of investigations on essential oils, aimed mainly at differentiating those from different plants and improving their quality. Distillation procedures, being particularly effective with essential oils, were prominent. From 1730 to 1732 he turned the traditional distillation procedures to analyze meat.[114] Returning to plants, he undertook once more a broad project to analyze them by the customary methods. "Having

114. C.J. Geoffroy, "Observations sur les huiles essentielles, et sur les differentes manières de les extraire et de les rectifier," *HAS*, 1721 [publ. 1723], 147–166; "Differens moyens d'enflamer, non-seulement les huiles essentielles, mais mesme les baumes naturels, par les esprits acides," ibid., 1726 [publ. 1728], 95–105; "Observations sur le méslange de quelques huiles essentielles avec l'esprit de vin," ibid., 1727 [publ. 1729], 114–120; "Suite d'observations sur les huiles essentielles, leur altération et la manière de rectifier celles de certains fruits," ibid., 1728 [publ. 1730], 88–99; "Examen chimique des viandes qu'on emploie ordinairement dans les boüillons," ibid., 1730 [publ. 1732], 217–232; "Suite de l'examen chimique des chairs des animaux," ibid., 1732 [publ. 1735], 17–30.

115. C.J. Geoffroy, "Manière de préparer les extraits de certaines plantes," *HAS*,

worked very much on many of these plants," he reported later;[115]

> The results presented me with nothing more remarkable than what we already have in the registers of the Academy, with the exception of a few differences that would not have rendered my detailed account very interesting. Therefore I did not believe that it was appropriate to publish observations little deserving to be in the annals of this Academy.

In this revealing acknowledgement of his disappointment, Geoffroy did not declare the old method useless or invalid. He gave it up because he could make with it no further progress. Merely confirming what Bourdelin and his colleagues had already observed half a century before was unattractive to academicians charged to bring before their colleagues what they had learned that was *new* to science. As he resumed his study of oils, Geoffroy turned more and more to extractions with water and spirit of wine. For a time he again fixed on narrowly practical goals, but increasingly he came to view plant matter as composed of the substances obtained by such methods. The essential salts, oils, and gummy and resinous parts were, he wrote, what one "ordinarily calls the principles of the plant, with the exception of the gross earth."[116]

The fate of the traditional analytical treatment as a foundation for the improvement or rational explanation of medical drugs is dramatically illustrated in a work of the elder Geoffroy. Among his multiple activities, Etienne Geoffroy published an influential pharmacopia, entitled *Treatise of medical materials, or the history of the virtues, choice, and usage of simple remedies.* A French translation of the volume dealing with exotic plants appeared in 1743. In the description of each plant, Geoffroy listed the quantities of each of the products obtained through a standard distillation analysis and commented on such matters as the particular qualities of those products, those in which the plant was especially abundant, and their medical uses.[117] In 1750 another volume on plants native to France appeared, written by an anonymous physician as a *Continuation of*

1738 [publ. 1740], 193–208, on 193.

116. Ibid., 196.

117. [Etienne] Geoffroy, *Traité de la matière médicale ou de l'histoire des vertus, du choix et de l'usage des remèdes simples, 2, Traité des végétaux,* sect. 1, *Des médicamens exotiques* (Paris, 1743).

the medical material of M. Geoffroy. In the advertisement, the author pointed out that details of the chemical analyses of the plants had been deleted, because "we have recognized by experience that the public draws no benefit from this detail. Who, in fact, except a professional chemist need trouble himself to know if a plant contains so much of oil, so much of phlegm, so much of salt, and so much of earth?" It is the properties of the plant that are of interest, "and nothing else." The author mentioned that Geoffroy himself had come to this point of view before his death.[118] The experience of Geoffroy's younger brother shows, however, as we have just seen, that professional chemists too had lost interest in these details. The old order was rapidly fading.

By this time the most popular teacher of chemistry in France was Guillaume-François Rouelle. Rouelle began giving private courses in pharmacy and chemistry sometime between 1737 and 1740, and public lectures at the Jardin du roi in 1742, when he succeeded Gilles Boulduc as demonstrator in chemistry there. Rouelle is best known for having popularized in France through his lectures the chemical doctrines of Stahl.[119] He was also much influenced by the writings of Boerhaave, Friedrich Hoffmann, and Caspar Neumann. In his treatment of plant chemistry, he presented what he clearly viewed as a new order of analysis based on extractions:[120]

> Chemistry finds marked differences in the products that it obtains on submitting the different parts of plants to analysis. It derives from their decomposition 1°, an aromatic spirit (in the case of odoriferous plants), which is ordinarily joined to an essential oil. 2°, an extractive material, which is soluble in water. 3°, a resin, which spirit of wine dissolves. 4°, a resino-extractive material, of which one part dissolves in water, and the other in spirit of wine, but the first is in greater proportion than the second. 5°, a matter analogous to the preceding, but in which the part soluble in spirit

118. [Anonymous], *Suite de la matière médical de M. Geoffroy,* vol. 1, sect. 2. *Des plantes de notre pays* (Paris, 1750), iv–v.

119. Rhoda Rappaport, "G.F. Rouelle: An eighteenth-century chemist and teacher," *Chymia, 6* (1960), 68–85.

120. Composite translation from "Cours de chimie, ou leçons de Monsieur Rouelle, recueillies pendant les années 1754, 1755 et rédigées en 1756. Revues, et corrigées, en 1757 et 1758," Manuscript in Science Library of Clifton College, Bristol, England, 78–79; and "Traité de chymie de Rouelle," manuscript No. 9 in Duveen-Lavoisier Collection, Cornell University Library, 57.

> of wine is predominant, M. Rouelle names it extracto-resinous. 6°, finally, a green matter that gives plants their color and that is found at their surface. This material is purely resinous and is not yet well known.
>
> It is by following that order of analysis that one attains a true idea of the composition of plants.

"The old chemists," according to Rouelle, were unaware of this kind of analysis, and confounded everything together through their practice of distilling plants over an open fire. The principles they obtained were in the plant, but the plant was not immediately composed of them, just as "one cannot say that the water and the earth that compose a brick are the materials of a house."[121]

In a discussion of this subject some time ago, I treated Rouelle's teaching as central to the emergence of solvent extractions as a new general order of analysis of plant matter, displacing the old order of distillation analysis.[122] Reinhard Löw has justifiably criticized my concentration on Rouelle as oriented too exclusively toward French chemistry. Löw has adduced two German chemists, Johann Friedrich Cartheuser and Rudolph August Vogel, who published "at the same time" as Rouelle "new principles" of plant composition similar to those Rouelle taught.[123] Simultaneity is, however, a concept difficult to apply historically to this case. Rouelle did not publish his views, and our knowledge of them derives mainly from lecture notes taken during the 1750s; but we do not know how much earlier Rouelle may have introduced these ideas into his courses.

In more basic terms, the question of priority is elusive because the "new order of analysis" was less a discrete investigative discovery than a pedagogical reorientation. Rouelle was recognizing and generalizing the outcome of a trend that had been evolving for a half century in the experimental exploration of plant chemistry. That is not to say that the reorientation was only a passive acknowledgment in the classroom of what had already happened in the laboratory. In France, at least, Rouelle's "new order" of plant analysis was a stimulus to productive investigative ventures during

121. Ibid.

122. Frederic L. Holmes, "Analysis by fire and solvent extractions: The metamorphosis of a tradition," *Isis, 62* (1971), 129–148, esp. 144–146.

123. Löw (ref. 108), 64–68.

the following decades. There is not time to discuss these investigations here, except to say that they contributed to the identification of new plant substances and toward the eventual merger of plant and animal chemistry into organic chemistry. Rouelle's teaching provides another strong example of what historians of science are beginning to notice in other areas: that pedagogical frameworks do not merely reflect completed research achievements. They can integrate and consolidate points of view that in turn shape further research activity.[124]

Simon Boulduc, Caspar Neumann, and Guillaume-François Rouelle each presented extraction analyses as a replacement for the traditional distillation analyses. The matter was not so straightforward, however, for, as their own discussions imply, the two methods operated on plant materials at different levels. One of Rouelle's early students, Gabriel François Venel, clarified the relation between these levels in a memoir that he communicated to the Academy in 1752. The extracts, resins, extracto-resins, resino-extracts, and other substances that Rouelle obtained with solvents Venel regarded as the "immediate materials...that constitute a plant chemically;" but chemists could not understand the nature of these materials themselves, he contended, unless they subjected them to an "ulterior analysis." Venel proposed to extend the method of solvents for this purpose. He also perceived, however, a useful role for the older distillation methods in this ulterior analysis. If distillation methods were applied to the simpler substances procured through extraction analyses, it would be possible, as it was not with the more complex plant matters, to infer from the nature of the distillation products their initial state in these substances.[125]

The question that chemists encountered from the seventeenth century onward in their analyses of plant matter—do the products of the analysis pre-exist in the plants themselves or are they artifacts of the analytical process?—identified a problem that has remained fundamental to all subsequent investigations of biological

124. See especially Kathryn M. Olesko, *Physics as a calling: Discipline and practice in the Königsberg seminar for physics* (Ithaca, 1990), and Gerald L. Geison, *Michael Foster and the Cambridge School of Physiology* (Princeton, 1978).

125. G.F. Venel, "Essai sur l'analyse végétaux," *Mémoires de mathématique et de physique présentées à l'Académie Royale des Sciences par divers savans*, 2 (1755), 319–332.

material. Eighteenth century chemists believed that the use of solvents instead of distillations could solve this problem. They regarded the products obtained in this way as "immediate," or "native" principles, even though they could not fulfill the accepted criterion for demonstrating chemical composition rigorously: that the original substance must be reconstituted by recombining what had been separated. Their reason for thinking so was partly based on the medical criterion that these substances often retained characteristic "virtues" or other properties of the plants from which they were drawn. Their confidence also derived from the traditional chemical view that the chief agent of material alterations was fire. By obviating the need for strong heating, the "new order of analysis" enabled them, they thought, to separate the constituents of plant matters without changing them. Even though their criterion was essentially negative, it did not, on the whole, seriously mislead them.

In this lecture I have outlined a second domain in which eighteenth century chemists carried on a progressive investigative tradition, and in which they reached a major conceptual and methodological transformation independent of the questions central to the chemical revolution. Someone might object that the story I have told is incomplete, because the resolution of the problem of "ulterior analysis" was only made possible by Lavoisier's new theory of combustion. It is true that we could construct a very satisfying story that would culminate in Lavoisier's elementary analyses of a number of plant substances. We could also, however, trace through the same period major developments in the identification of "immediate principles" of plant and animal matter in which neither Lavoisier nor his methods and theories played an essential part. Both lines of investigation were fundamental to the later growth of organic chemistry. Chemical investigation, like any other human activity, is an ongoing stream, and I believe, along with Hayden White and others, that historians create useful artifacts when they separate from such activity the plots of stories with discrete beginnings and endings.[126]

126. Hayden White, *Tropics of discourse: Essays in cultural criticism* (Baltimore, 1978).

4 CHEMISTRY AND INDUSTRY

The eighteenth century was pivotal to the most pervasive transformation that Western society has undergone: its industrialization. An intense historical debate has raged over whether or not the formative stage of industrialization was significantly dependent on science. The polar positions on this question are as well known as they are inadequate: on the one hand, that the scientific revolution of the seventeenth century was in some global way the essential foundation for the industrial revolution of the eighteenth century; on the other hand, that the industrial revolution was achieved through the inventions of empirical craftsmen little acquainted with or influenced by the great intellectual events out of which modern science developed.

The industrial revolution is identified most prominently with coal and iron, textile manufacturing, machinery substituted for hand labor and driven eventually by the motive power of steam, and the factory organization of production. It is recognized, however, that the emergence in the late-eighteenth and early-nineteenth century of a heavy chemical industry, producing acids, alkalis, and other basic substances on a large scale, was also critical to the dynamic movement of industrialization. Although the chemical side of the industrial revolution has not been studied as extensively as processes that were mechanized, several comprehensive historical studies exist. The pioneering book, *The chemical revolution* by Archibald and Nan Clow, traced the development in Scotland of the manufacture of salts, alkalis, vitriolic acid, glass and pottery, the transition from processes employing wood for heat to those relying on coal, the formation of the bleaching industry, and other manufactures based on chemical operations.[127] More recently, John

127. Archibald Clow and Nan Clow, *The chemical revolution: A contribution to social technology* (London, 1952).

Graham Smith has written a fine account of *The origins and early development of the heavy chemical industry in France.*[128] A.E. Musson and Eric Robinson's *Science and technology in the industrial revolution* also devotes extensive attention to the role of chemistry.[129] Beyond these broad treatments there have recently appeared a number of suggestive articles giving case studies of individuals who applied chemical methods for industrial or other commercial purposes.[130]

1. RELATIONS BETWEEN SCIENCE AND INDUSTRY

The question that has been asked repeatedly about the industrial revolution as a whole arises with particular intensity in these historical studies of chemical industry. Were the early large-scale commercial chemical operations based upon applications of a science of chemistry, or were they outcomes of trial-and-error adaptations of small-scale craft processes to industrial conditions?

An examination of recent historical studies of this question suggests that answers depend as much upon historians' definitions of science and the requirements they invoke for a demonstration of a meaningful impact of science on technology as they do upon the historical evidence that can be brought to bear. According to Rupert Hall, the stringent standard set by some historians is that to show strong connections one must identify "precise examples of a technical innovation's being derived consciously from pre-existent theoretical knowledge of a non-trivial character."[131] Charles Gillispie

128. John Graham Smith, *The origins and early development of the heavy chemical industry in France* (Oxford, 1979).

129. A.E. Musson and Eric Robinson, *Science and technology in the industrial revolution* (Manchester, 1969).

130. Examples are Charles C. Gillispie, "The discovery of the Leblanc process," *Isis, 48* (1957), 152–170, and "The natural history of industry," ibid., 398–407; Neil McKendrick, "The role of science in the industrial revolution: A study of Josiah Wedgwood as a scientist and industrial chemist," in *Changing perspectives in the history of science: Essays in honour of Joseph Needham*, ed. Mikulas Teich and Robert Young (London, 1973), 274–319; and Homer E. Le Grand, "Theory and application: The early chemical work of J.A.C. Chaptal," *British journal for the history of science, 17* (1984), 31–46.

131. A.R. Hall, "What did the industrial revolution in Britain owe to science," in *Historical perspectives: Studies in English thought and society in honour of J.H. Plumb*, ed. N. McKendrick (London, 1974), 129–151, on 129.

adopted such a standard in his study of the Leblanc soda process. To prove that "science revolutionized manufacturing" in some area, one must "trace the course of [a] significant theoretical concept from abstract formulation to actual use in industrial operations."[132] Musson and Robinson base their evaluation of the connection on a less theoretical view of science. "Modern science is a combination of theory and experiment; it has been developed, not in ivory towers, but in laboratories....If by empiricism we mean procedure by trial and error, then the progress of science has been empirical."[133] Concomitantly, they allow that industrial processes reached by such procedures can be regarded as derived from science. The criterion for which they seek evidence is that those who developed industrial operations were familiar with contemporary science and were connected through social networks with scientists.

Such discordant general criteria derive, I believe, from unanalyzed assumptions in what Thomas Kuhn has identified as the historiographic tradition in which "the body of scientific knowledge [is] conceived as one;" and in which the methods, the philosophical climate, or the social organization characteristic of an age are expected to apply to all areas of the study of nature. Such an approach has dominated the treatment of the seventeenth and eighteenth centuries, in which the revolution in astronomy and mechanics culminating with Newton has been viewed as paradigmatic for the emergence of modern sciences in general. In contradistinction to this view, Kuhn stressed in his brilliant essay on the "Mathematical versus experimental traditions in the development of physical science," how different the patterns of development have been in two contrasting clusters of fields. The "classical sciences," including astronomy, geometrical optics, and statics, are "the only parts of physical science which during antiquity became objects of" specialized research traditions, and which were treated mathematically. In the early stages of the scientific revolution, the study of motion was added to this list. Each of them then underwent a profound intellectual transformation. Although important experiments were carried out within these fields, they were few in number and designed to test consequences foreseen through theory. Throughout

132. Gillispie, "Natural history" (ref. 130), 398–399.
133. Musson and Robinson (ref. 129), 3–4, 10–59.

the eighteenth century these fields continued to be developed largely through mathematical analysis. The second cluster of sciences, termed Baconian by Kuhn, dealt with phenomena such as heat, magnetism, and electricity, which had not previously been subjects of sustained investigation. They were rooted in craft practices, were empirical, emphasized careful observation and new forms of instrumentation, and lacked tightly structured theoretical foundations. Since they were only emerging as distinct areas of inquiry in the seventeenth century, they did not undergo major conceptual reconstructions.[134]

The criteria set by Hall and by Gillispie for establishing causal connections between science and industrialization appear to reflect that historiographic tradition in which the theoretical transmutations that occurred within Kuhn's "classical sciences" are taken as models for the formation of all modern science. A development leading ultimately to industrial processes should originate in a theoretical concept, because it was the theoretical restructuring that took place within the century preceding the industrial revolution that was central to these fields. Musson and Robinson's image of scientific development appears, in contrast, to reflect patterns of development more typical of Kuhn's "Baconian" sciences.

"Chemistry," Kuhn concedes, "presents a case of a different and far more complex sort," not fitting entirely into either of his categories. Many of its methods and operations "had been developed long before the scientific revolution;" but these had been pursued primarily by craftsmen and alchemists before a re-evaluation of the techniques involved incorporated chemistry into "the experimental search for knowledge."[135] The special position of chemistry as a subject rooted in craft, yet endowed by the eighteenth century with a longer conceptual development than the other "Baconian fields"—and with aspirations, as mentioned in my first lecture, to distance itself from its crafts roots in order to attain the status of an academic science—greatly complicates questions about the relation between science and the development of a chemical industry. Because the investigative chemists of the eighteenth

134. Thomas S. Kuhn, "Mathematical versus experimental traditions in the development of physical science," in *The essential tension* (Chicago, 1977), 31–65.

135. Ibid., 46–47.

century occupied themselves extensively with some of the same traditional craft processes whose transformation into large-scale manufacturing operations underlay the chemical industrial revolution, it becomes a subtle matter to evaluate the relative degree to which the traditional practices and the investigative science contributed to the industrial developments.

The issue of the historical relation between science and the chemical industry has also been colored by the canonical story of eighteenth century chemistry that I discussed in the first lecture. Historians easily assume that a science adequate to serve as a theoretical foundation for industrial application came into being only with the chemical revolution. Thus Gillispie took as a mark of the absence of significant chemical theory in the development of the industrially crucial Leblanc soda process the fact that "Lavoisier is never even mentioned."[136] In an otherwise cogent discussion of the prominent industrial chemist Jean Antoine Claude Chaptal, Homer Le Grand tends to merge the question of whether Chaptal applied chemical theory meaningfully to his studies of chemical technology with the question of whether he was an early and unambiguous supporter of Lavoisier's system of chemistry.[137]

On the basis of his case study of the Leblanc process, Gillispie proposed that chemical theory was not applied during the eighteenth century to change industrial processes, but to explain processes already employed, as part of a broader effort to incorporate the arts into the structure of human knowledge.[138] Even if we accept this view, there are historical pitfalls in our path. According to Gillispie, eighteenth century chemistry was inadequate even to explain the Leblanc process. Neither Leblanc nor his contemporaries undertook a full analysis of the process he had invented, and the reaction was not "properly described until near the end of the nineteenth century, some time after the Leblanc process had already begun to go out of use."[139] John Smith judges in a similar way that eighteenth century chemistry could not provide a

136. Gillispie, "Natural history" (ref. 130), 404.

137. Le Grand (ref. 130), 31–46.

138. Gillispie, "Natural history" (ref. 130), 404–405.

139. Gillispie, "Leblanc process" (ref. 130), 170. A.R. Hall (ref. 131), 138–139, follows Gillispie's interpretation judging that chemical science offered little assistance toward the development of commercial manufacturing.

correct interpretation of the combustion process through which vitriolic acid was produced commercially from sulfur. Not until 1806 did Clément and Desormes achieve a "true explanation" of the circumstances found empirically to be essential to the success of the process.[140] "True" explanations, however, are never the whole truth, and the understanding of chemical reactions in any era is less complete than in succeeding eras. We should ask not whether eighteenth century conceptions of the chemical processes central to early industrialization were correct in the light of later knowledge, but whether in their own time they contributed to the large-scale realization of these processes.

I have not myself pursued original historical studies of eighteenth century industrial chemistry. What follows, therefore, will be only a suggestion that the point of view concerning the chemistry of salts developed in my second lecture can provide fresh perspectives regarding the relation between science and the industrial revolution. I would like to illustrate through two examples the complexity of the reciprocal interactions between eighteenth century chemical investigation and commercial processes, and the diversity of the patterns discernible in individual cases.

2. THE INDUSTRIAL PRODUCTION OF VITRIOLIC ACID

As Clow and Clow and John Smith have both emphasized, vitriolic acid was the first chemical to be manufactured on a large scale. The rising levels of its production during the late eighteenth and early nineteenth century can be viewed as a prime indicator of the growth of a heavy chemical industry.[141] Long before that the substance had been a key reagent in the repertoire of chemists, metallurgists, and pharmacists. During the seventeenth century chemists obtained vitriolic acid, known then more commonly as spirit of vitriol or, in more concentrated form, as oil of vitriol, by two methods. One way was to distil green vitriol, a mineral containing iron. The name vitriolic acid thus derived, as many early chemical names did, from one of its principal sources. The other

140. Smith (ref. 128), 57, 63.
141. Clow and Clow (ref. 127), 130; Smith (ref. 128), 4.

method for procuring it was to burn sulfur in a closed vessel.

In his popular seventeenth century textbook, *Cours de chymie*, Nicolas Lemery described both methods. The product of the distillation of green vitriol he called spirit of vitriol. That of the burning of sulfur he called spirit of sulfur; but he commented that the latter "is nothing but a resolved vitriolic salt that does not differ at all from spirit of vitriol, except perhaps in its taste, which is not so empyreumatic, because it has not received the violence of so great a fire."[142] The identification with one another of similar substances obtained from different sources was almost always problematic during the seventeenth century.

Lemery noted that "a great number of machines have been invented to draw out the spirit of sulfur." The most common was a glass bell jar under which the sulfur was burned, and against the walls of which the spirit of sulfur formed. He himself employed an earthen vessel with an elongated neck and a small hole to allow the access of air. He placed melted sulfur in a bowl supported on a second inverted bowl, under which the acid spirit collected.[143]

Seventeenth century chemists understood the four known "species" of vitriol to be composed of "an acid spirit" combined respectively with iron, copper, or mixtures of both metals. In discussing the distillation of green vitriol, Lemery remarked that:[144]

> Acid spirits are nothing but salts rendered fluid by the force of the fire that disengages them from their most earthy part... One can restore body to them by pouring them on any alkali. For example, the spirit of vitriol, having remained for some little time on iron, is reincorporated into vitriol, and spirit of nitre, poured on salt of tartar, forms a saltpeter.

As mentioned above, Lemery noted that the spirit of sulfur obtained by burning sulfur was also a vitriolic salt; but he did not further relate the composition of the sulfur from which it had come to the composition of the vitriols.

The idea of composition that Lemery applied here to a small number of examples was consolidated and generalized by the beginning of the eighteenth century in the conception of middle

142. N. Lemery, *Cours de chymie*, 4th ed. (Paris, 1681), 535.
143. Ibid., 533–534.
144. Ibid., 509.

salts as a broad class of substances divisible into distinct categories, one of which contained acids combined with metals as in green vitriol.

During the first half of the eighteenth century knowledge of the chemistry of vitriolic acid was greatly expanded. By the time that Macquer summarized the subject in his *Dictionnaire de chymie* in 1766, there were well characterized neutral salts, obtainable in crystalline form, for combinations of the acid with most of the common metals; and its combinations with recently discovered metals such as bismuth, zinc, and cobalt were under examination. The distinction made by Duhamel between vegetable and mineral alkali had enabled a clear separation of the respective salts of these two alkalis with vitriolic acid. The identification of distinct alkaline earths clarified the composition of selenite and of alum as combinations of vitriolic acid with calcareous and argillaceous earth, respectively. The characterization of so many of its salts also made the identification of vitriolic acid itself more precise, so that variations in properties such as the one described earlier by Lemery could be attributed with confidence to impurities. The order of replacement of each of these bases in combination with vitriolic acid, and of vitriolic acid among the acids combining with each class of bases, were prominent in Geoffroy's table of *rapports* and its expanded successors.[145]

The main contribution of Stahl's phlogiston theory to the chemistry of vitriolic acid was to explain its relationship to the sulfur from which it was commonly produced. According to that theory, sulfur was a combination of vitriolic acid with the inflammable principle, and when sulfur burned in the presence of air, in such a way that its phlogiston was fully dissipated, only vitriolic acid remained.[146]

Stahl did far more, however, than merely to provide a theoretical rationale for a previously known chemical operation. By mixing a salt of vitriolic acid with charcoal and heating them, he was able to reconstitute sulfur. Since charcoal was also a prime source of phlogiston, it was natural to interpret this process as a transfer of its inflammable principle to the vitriolic acid, their union producing the sulfur. Moreover, he incorporated into this scheme another acid,

145. [Macquer] (ref. 35), *1*, 23–31; *2*, 430–431.
146. Ibid., *2*, 494–495.

volatile sulfurous acid, a substance known since the seventeenth century to form when sulfur was heated less strongly than required to produce vitriolic acid. Stahl explained volatile sulfurous acid as the result of the partial removal of phlogiston from sulfur. He could also produce it from vitriolic acid by processes understood to combine the latter with less phlogiston than was necessary to form sulfur.[147]

There were further ramifications of this network of interrelated substances and operations connected with vitriolic acid and the sources from which it was ordinarily obtained. The preceding is enough, however, to indicate that in the mid-eighteenth century there was an extensive knowledge of its properties and its reactions, based on the cumulative results of numerous experimental investigations and organized within the coherent conceptual frameworks of the neutral salt and the phlogiston theory.

The fact that it was also in the mid-eighteenth century that vitriolic acid began to be manufactured "en grande," as the French chemists put it, might lead us to expect that the contemporary state of knowledge about its chemistry must have been pertinent to this dramatic commercial development. From the accounts given by Clow and Clow and by John Smith, however, this appears not to have been the case. The first person to produce the acid on a large scale was Joshua Ward, described as a quack doctor who stayed for a time in France to "evade justice in connexion with a parliamentary offence" in England, returned in 1733, and set up three years later a works in Twickenham to produce "oil of vitriol by the bell," that is, by burning sulfur under a bell jar. Apparently Ward or his partner John White succeeded simply by scaling up the apparatus by which this process had been carried out since the seventeenth century. He employed glass vessels of forty to fifty gallons capacity, and was able to reduce the price of a pound of vitriolic acid to about that for which an ounce had formerly sold. A critical factor in his success was that he added a small amount of nitre (that is the salt composed of nitrous acid and fixed vegetable alkali) to the sulfur. This was not a new innovation. Already in the late seventeenth century some chemists had found that nitre increased the yield; but others, including Lemery, rejected it because it also

147. Ibid., 2, 503–507.

produced some nitrous acid mixed with the vitriolic acid.[148]

Further increases in the scale of the operation during the eighteenth century, making it a truly industrial process, depended mainly on technical changes. To escape the limit on the size to which glass vessels could be blown, the second major manufacturers of vitriolic acid in Britain, John Roebuck and Samuel Garbett, substituted lead chambers in which to carry out the process. In doing so they relied on a particular chemical property of vitriolic acid, that it did not attack lead; but this was based on an old observation, made by Glauber in the seventeenth century.[149] Thus nothing essential to the adaptation of vitriolic acid production to large-scale manufacturing conditions appears from these historical accounts to have been dependent upon the outcome of recent chemical investigations.

Smith, who has described in detail the efforts of chemists in France late in the eighteenth century to improve the efficiency of the process, maintains that they were prevented from doing so by a "network of largely erroneous ideas" concerning its chemistry. These centered around their failure to understand the role of the nitre, leading them to misguided attempts to eliminate it in order to save its cost.[150] It is true that the conceptual structure of eighteenth century chemistry provided no explanation for the effect of the nitre that appears in retrospect convincing. Macquer classified the effect as an example of a "clyssus." "One calls clyssus," he wrote in his dictionary under that heading, "the vapours which are exhaled during the detonation of nitre with any inflammable body." Reporting that "An English chemist has some time ago found a means to obtain vitriolic acid from sulfur, with great profit and advantage, by burning it in closed vessels with the addition of a very small quantity of nitre," Macquer viewed the commercial success of the operation as an example of the utility of the clyssus: "for the vitriolic acid thus extracted with the aid of the detonation of a little nitre in the closed vessel must be regarded as a true clyssus of sulfur."[151] This explanation may appear to us not only outmoded, but devoid

148. Clow and Clow (ref. 127), 131–132; Smith (ref. 128), 5–6.
149. Clow and Clow (ref. 127), 133–134.
150. Smith (ref. 128), 57–63.
151. [Macquer] (ref. 35), *1*, 266–268.

of heuristic value for improving the process. We should remember, however, that chemists at all times have had to supplement conceptual structures that predict incompletely the circumstances required to carry out a given process with trial-and-error searches for "optimal conditions." The idea of a clyssus at least entrenched the utility of the nitre for Macquer within a class of analogous effects. If eighteenth century theories of combustion were inadequate fully to refine the process by which vitriolic acid was produced, the phlogiston theory provided at least a framework within which to understand its general nature, as did the oxygen theory a little later.

The industrial production of vitriolic acid in the eighteenth century can thus not be attributed to a direct application of current chemical theory or new experimental developments. Theoretical understanding of the combustion process was clearly not essential in order to carry it out on a large scale, because the industrial process was derived from a small scale one that had been in use long before a theory of combustion was available to explain it. Derek Price has argued that technological innovations are most often connected, not to the most recent "research-front" science, but to older established scientific knowledge.[152] In this case the technology originated from a stage in the emergence of chemistry during the seventeenth century in which it is difficult to distinguish "scientific" procedures from craft practices.

To conclude that the origin of the large-scale production of vitriolic acid was not directly dependent upon eighteenth century advances in the science of chemistry is not to exclude more subtle impacts upon the development of the process. The ever greater scope of eighteenth century analytical procedures, the increasing assurance with which vitriolic acid and related substances could be separated and identified, their interactions traced, and their impurities detected, must surely have aided industrial chemists who sought to evaluate or improve the commercial output of their product. Moreover, some of the uses to which the ever increasing quantities of vitriolic acid manufactured could be put were linked to the growing knowledge of chemistry. The earliest large-scale uses

152. Derek J. DeSolla Price, "Is technology historically independent of science? A study in statistical historiography," *Technology and culture*, 6 (1965), 553–568, esp. 564–568.

for the acid—in hat-making, the preparation of leather, dyeing, and cotton printing—may have been largely empirical.[153] By the end of the century, however, the production of vitriolic acid was closely tied to that of another chemical whose connection with eighteenth century investigations of the chemistry of salts is unmistakable.

3. THE MANUFACTURE OF SODA

"The roots of the soda industry," according to John Smith, "are to be traced to an important scientific study of the chemical constitution of common salt which Duhamel du Monceau presented to the Academy of Sciences in 1737." This investigation "laid the scientific foundations for the manufacture of soda, [by] demonstrating on a small scale the practical possibility of producing soda from salt."[154] As I pointed out in my second lecture, that was not Duhamel's own immediate objective. He aimed instead to distinguish the alkali contained in sea salt from the ordinary fixed alkali obtainable from plant matter. In order to do so, he devised a strategy to separate the alkali from its acid, after which he could show both that the properties of the two alkalis were different and that the salts formed by the alkali from sea salt with each of the common acids differed from the corresponding salts of ordinary fixed alkali. In my discussion of Duhamel's study I presented it not as an isolated event, but as a highlight of the ongoing investigative enterprise in chemistry within the Academy oriented around the classificatory conception of the middle salt and guided in part by Geoffroy's table of *rapports*. The ramifications of Duhamel's work for the production of soda thus constitute a clear-cut channel leading from a place where the ethos of experimentation in the interest of scientific advancement prevailed toward the entrepreneurial world of industrial chemistry.

As Smith has stressed, those who took an interest in the commercial possibilities inherent in Duhamel's studies derived from it more than the idea that fixed mineral alkali could be obtained from common salt, making available the abundance of sea water as a

153. Smith (ref. 128), 34–36.
154. Ibid., 194–195.

source. Some of them also attempted to devise manufacturing methods by modifying the experimental strategy he had used. The first step was to displace the marine acid with vitriolic acid, producing what was then known as Glauber's salt. They followed Duhamel in reducing the vitriolic acid in that salt to sulfur by heating it with charcoal, a reaction they too would have interpreted as imparting the inflammable principle to form the sulfur. Duhamel had next displaced the sulfur with acid of vinegar (or acetous acid) which was in turn removable by distillation to leave the alkali. The cost of the acetous acid, however, made these last steps impracticable for large-scale production. They therefore sought alternative methods to separate the sulfur from the alkali.[155]

As Smith mentions, and Musson and Robinson have described in further detail, the first sustained effort to obtain alkali from sea water for commercial purposes was carried out between 1766 and 1771 by James Watt and James Keir. Following the advice of the distinguished chemist Joseph Black, they did not adopt Duhamel's approach, but attempted to make "alkaline salt from common salt and lime" by allowing the former to stand for long periods of time on cakes of the latter. For reasons that are not entirely clear, Watt eventually grew discouraged, and although the process did produce some alkali, nothing further came of their experiments. In Britain, sea-side plants, a traditional natural source of the alkali that Duhamel distinguished from the alkali traditionally obtained from wood ashes, were still available readily enough so that there was no strong incentive to produce the alkali by other means.[156] In France, on the other hand, disruptions of the supply of potash from North America and the difficulties obtaining the Spanish *barilla* plant that served as France's only major source of the mineral alkali so heightened interest in alternatives that in 1783 the Academy of Sciences offered a prize for the one who could "find the simplest and most economical process" to extract from marine salt "the alkali that serves as its base, in its pure state, free of all acid or other combination."[157]

155. Ibid., 195–198.

156. Ibid., 195; Musson and Robinson (ref. 129), 352–371; Eric Robinson and Douglas McKie, eds., *Partners in science: Letters of James Watt and Joseph Black* (London, 1970), 16–26.

157. Smith (ref. 128), 198–199.

Both before and after the announcement of this prize, an impressive number of French chemists and their associates made attempts to develop a practical means to the end it promoted. They included such able scientists as Chaptal and Guyton de Morveau. A report on these processes made during the revolutionary period to the Committee of Public Safety and published afterward in the *Annales de chimie* makes clear that several of these competing groups applied current chemical knowledge of acids and bases and of their orders of affinity resourcefully to reach their practical objective. A number of them did succeed in producing substantial quantities of fixed mineral alkali—or as it came to be known after the new nomenclature appeared, sodium carbonate. Most of them, however, encountered technical, financial, or other difficulties that prevented them from attaining commercial success.[158]

As is very well-known, the process that ultimately resolved these practical problems and dominated the rise of the heavy alkali industry was invented in 1789 by Nicolas Leblanc, a minor chemist compared to some of his competitors. The first step in the Leblanc process, like several of the others, was a large-scale version of Duhamel's laboratory procedure: the displacement of marine acid (by then called muriatic acid) by vitriolic acid (by then called sulfuric acid). The crucial and innovative second step, however, heating the resulting sodium sulfate with a mixture of charcoal and lime, has appeared to various commentators to be an empirical move, made without logical foundations in the existing chemistry of salts or in Lavoisier's new theory of combustion.

It is the apparent absence of a compelling scientific basis for this most original aspect of the Leblanc process—the main source of its commercial success—as well as Leblanc's minimal achievements as an investigative chemist that prompted Gillispie to term the Leblanc process an invention that neither resulted from the application of science nor was adequately explained by the science of the time. He further argued from this crucial case against the view that there were strong causal connections between science and the industrial revolution.

158. Ibid., 196–209; Lelievre, Pelletier, D'Arcet, and Alexandre Giroud, "Extrait d'un rapport sur les divers moyens d'extraire avec avantage la soude du sel marin," *Annales de chimie*, 19 (1797), 58–156.

It may well be, however, that a plausible connection can be found between Leblanc's innovative second step and the structure of eighteenth century chemistry. Leblanc himself later testified that his initial inspiration had come from an article published in 1789 in the journal *Observations sur la physique.* Reporting on various observations he had made during a trip to England, the editor of that journal, J. C. de la Métherie, described a number of attempts that various individuals had made to decompose marine salt so as to obtain its acid and its alkali separately. None of them had been successful in his opinion, yet to achieve this end, he thought, "is of the greatest importance because of the continual use made of these two substances in the arts." After mentioning that several persons had tried to decompose marine salt by means of quicklime, and that some had been "fortunate, but this process is not known," de la Métherie added a proposal of his own:[159]

> There is a way to attain that decomposition that would be very certain, but it would perhaps be too expensive. This would be to pour vitriolic acid on marine salt: the marine acid would be disengaged and pass into the vessels, and the residue would be sodium vitriol, or Glauber's salt. One would next decompose the sodium vitriol by calcining it with charcoal. The vitriolic acid will be disengaged in the form of sulfurous acid, and the sodia will remain pure.

As can be seen, the operations proposed were the same as the first two steps in Duhamel's experimental procedure; but de la Métherie claimed that the product would be sulfurous acid rather than sulfur. Gillispie rejected the possibility that this proposal could have provided the clue that led to the Leblanc process, because "the reaction described does not work; its product will be not soda, but sodium sulfide." In its place, Gillispie proposed that Leblanc may have obtained his suggestion from a preceding section in de la Métherie's article describing the use of charcoal and limestone in the smelting of iron. Drawing what Gillispie calls a "fallacious analogy," Leblanc tried charcoal and limestone as a means to remove vitriolic acid from the alkali. If so, it was a fortuitous juxtaposition of topics in de la Métherie's article rather than a coherent

159. J.C. de la Métherie, "Discours préliminaire," *Observations sur la physique*, 34 (1789), 39–46, on 44.

hypothesis that prompted Leblanc to take the critical step toward his invention.[160]

John Smith, on the other hand, defends the view that de la Métherie's proposal probably formed the starting point for Leblanc's experimental endeavors. Supplementing Leblanc's testimony with an account of the discovery that Leblanc's former associate M. J. Dize gave after Leblanc's death, Smith reconstructs the investigation as beginning with an effort to carry out de la Métherie's process. The product did turn out to be a combination of the alkali with sulfur rather than with sulfurous acid. Setting out then to find a way to separate the sulfur from the alkali, Leblanc and Dize succeeded first by directing a stream of carbonic acid gas over the salt. Afterward they were able to do so by fusing a solution of the salt with charcoal and carbonate of lime, and finally they reached the same end by adding the charcoal and lime directly to the sodium sulfate.[161]

Even while accepting Leblanc's testimony that he had been inspired by de la Métherie's proposal, Smith called the latter "apparently just a speculative suggestion, perhaps deriving from something he had heard in England."[162] Within the context of eighteenth century chemistry, however, the proposal can be viewed as a serious hypothesis. As a late adherent of the phlogiston theory, de la Métherie probably had in mind Stahl's discovery that vitriolic acid can be converted to volatile sulfurous acid by transferring to it some of the inflammable principle but not enough to form sulfur. Because of the volatility of sulfurous acid, it could then be expected, unlike vitriolic acid, to separate readily from the alkali. The proposal thus constituted another tactical maneuver to accomplish what Duhamel had done by reducing the vitriolic acid all the way to sulfur and then displacing the latter with the readily removable acetous acid.

The process did not work out as de la Métherie predicted. Leblanc was forced to search out another means to get rid of the sulfur he had produced instead of sulfurous acid. That outcome should not prevent us from defining Leblanc's use of the proposal

160. Gillispie, "Leblanc process" (ref. 130), 167–170.
161. Smith (ref. 128), 209–215.
162. Ibid., 210.

as a test of a scientific hypothesis. The history of science is replete with discoveries resulting from the pursuit of hypotheses that were themselves falsified along the way.

Whether we attribute critical steps in the fine structure of Leblanc's creative experimental investigation to de la Métherie's suggestion, to his own theoretical insight, or to a fallacious analogy, the enormously successful industrial process resulting eventually from his labors remains linked firmly to prior scientific discoveries; for the whole late eighteenth century effort to produce soda on a large scale fitted within the conceptual and experimental framework set by Duhamel's investigation of the alkali in sea salt. Gillispie's view that the Leblanc process did not arise from an application of science to industrial operations is a reflection on the dominant historiographical traditions within which he wrote thirty years ago. Viewing the history of science as an intellectual process, he sought an influence leading from "abstract theory" to industrial practice, and found none. Even if affinity theory were found to have played a part he would not be impressed, because "the theory of affinities was devoid of abstract interest."[163] Assuming the canonical view that chemistry emerged as a theoretical science only with Lavoisier's theory of combustion, Gillispie looked for an influence of that theory on Leblanc and also found nothing. That he did not look for an application of a scientific discovery made fifty years earlier by a chemist in the Academy of Sciences reflects the scant attention that historians of science have paid to the investigative activity of early eighteenth century chemistry. John Smith did reach back to Duhamel as the starting point for the later industrial chemists, but he did not place Duhamel's work into its contemporary scientific context, probably because a historical approach to eighteenth century chemistry fixed on the drama of the chemical revolution has found little room to portray that context.

In the two cases I have discussed—the transitions to large scale production of vitriolic acid and of soda—we can see illustrated quite different relations between science and industrial development. In one, the interactions were diffuse and the impact of recent scientific progress may have been marginal. In the other, a specific scientific investigation played a central role, in a sense made possible the

163. Gillispie, "Natural history" (ref. 130), 404.

subsequent industrial change. Were we to explore other processes important during the eighteenth century, such as glass and soap-making, or the manufacture of alum, we would find patterns of interactions between science and manufacturing that differ from either of the preceding cases.[164] We should not rush to generalize from one or two instances or attempt too soon to differentiate essential connections from contingent circumstances.

In fixing my attention during this lecture on connections between science and the growth of a chemical industry, I do not purport to offer a balanced picture of the factors conditioning that growth. Obviously entrepreneurial opportunities and limitations, demand for products, and broader socio-economic conditions were all central to these developments. Clow and Clow, Smith, and others have described these aspects of the emergence of a chemical industry in considerable detail, and my objective is only to connect some of the stories they have told to a story of eighteenth century chemical science that has been less adequately told.

The question of what part science played in the origins of an industrial revolution that has thoroughly transformed the economy, the geography, the social structure, and the culture of most of the world is surely one of the broadest issues that can be raised concerning science and its social context. The preceding examples imply that this issue cannot be understood historically without close attention to the conceptual structures, the operational methods, and specific investigational episodes embedded within the development of the pertinent sciences. What more persuasive argument can there be that an illuminating external approach to the history of science must be symbiotic to finely structured studies of what is sometimes called the "narrowly internalistic" dynamic of scientific investigation?

164. Robert P. Multhauf explores another important example in "Sal ammoniac: A case history in industrialization," *Technology and culture,* 4 (1965), 569–586. Multhauf is cautious about attributing a significant role to science.

5 THE CHEMICAL REVOLUTION

A final lecture devoted to the chemical revolution may appear to undermine the position I have maintained in the preceding lectures. Does this order not reinforce the conventional pattern that all treatments of eighteenth century chemistry culminate there? My purpose, however, has not been to displace the chemical revolution from its central role in the last act of the drama of eighteenth century chemistry. If the view that I have developed, that a structured discipline with a thriving investigative tradition existed in chemistry long before the revolutionary period is correct, then we must resituate and reconsider the meaning of that revolution.

1. THE HISTORICAL PLACE OF THE REVOLUTION

Scholars are currently reassessing the chemical revolution and its historical place. In a stimulating new article on "Lavoisier and the origins of modern chemistry," Arthur Donovan asserts that studies of Lavoisier and the revolution he led can be divided into two types: detailed reconstructions of Lavoisier's career and the revolution itself, and "attempts, suggestive at best, to locate Lavoisier's achievement in the larger context of the development of science in the latter half of the eighteenth century. The fact that few links have been forged to connect these two types of research stands as a serious inadequacy in our contemporary understanding of Lavoisier."[165] To this cogent remark I would add that it stands as an equally serious inadequacy in our contemporary understanding of

165. Arthur Donovan, "Lavoisier and the origins of modern chemistry," *Osiris*, 4 (1988), 214–231, on 214.

eighteenth century chemistry.

Reviewing earlier evaluations of Lavoisier's achievement, Donovan points out that Lavoisier has sometimes been seen as having reconstructed an existing science and sometimes as having transformed into a science what had previously been only a practical art. Favoring the second of these alternatives, Donovan argues that Lavoisier himself believed that he had brought chemistry "*into* science, the relevant model for science being provided by experimental physics." Donovan supports this view with several strong passages from Lavoisier's writings, including the celebrated prediction in his laboratory notebook in 1773 that his observations seemed destined "to bring about a revolution in physics and chemistry," and a letter to Benjamin Franklin in 1790 claiming that through his work "chemistry has been brought much closer than heretofore to experimental physics."[166]

Evan Melhado has recently made the more radical claim that Lavoisier was not really a chemist, but a physicist who exploited certain chemical phenomena as convenient means for exploring the nature of heat. From a rather elaborate argument beginning with a definition by the French Stahlian chemist Gabriel François Venel of heat in the free state as part of physics, and in its role in combinations as part of chemistry, Melhado concludes that Lavoisier's ideas on the nature of heat fell primarily in the domain of physics. "Lavoisier is," according to Melhado, "more readily lodged in physics than in chemistry." Therefore, "If Lavoisier transformed chemistry, he did so from without."[167]

Like some previous efforts to isolate a central thread woven through Lavoisier's scientific aims or achievements, these interpretations generalize broadly from programmatic statements that Lavoisier made within specific, localized contexts; and they stress certain aspects at the expense of others that were to Lavoisier equally important. Melhado summarizes recent debates among historians about whether Lavoisier was concerned primarily with processes that fix or release air and only secondarily with the nature of heat; or whether his interest in heat was primary and he took up

166. Ibid., 215–231.

167. Evan M. Melhado, "Chemistry, physics, and the chemical revolution," *Isis*, *76* (1985), 195–211.

combustion only because it was a process involving the release of heat.[168] Such issues are unresolvable, because Lavoisier's thought was too complex and flexible to be reduced to lapidary formulas or priority gradients. If one follows any strand of his investigative pathway closely, one finds that means and ends for him were often interchanged, that he viewed the same clusters of problems from different perspectives at different times. It is risky to assign fixed primary and subordinate places for his range of interests. The renewed interest in experimental physics as the model for Lavoisier's chemistry has been engendered in part by an important article Henry Guerlac wrote a decade ago. Guerlac showed how the experiments that Lavoisier pursued with Laplace between 1776 and 1783, which led to the rigorous definition and quantification of what had hitherto been Lavoisier's rather loose ideas on the nature of heat, added both a style of reasoning characteristic of mathematical physics and an experimental program lying on the borderland of physics and chemistry to Lavoisier's scientific trajectory.[169] This joint effort was, however, only one facet of a broader research program that Lavoisier sustained both before and after his collaboration with Laplace. Guerlac did not demonstrate that Lavoisier's investigative enterprise as a whole was through this association subsumed within physics.

Lavoisier admired not only experimental physics, but also mathematics; and he sometimes wrote that he aimed to introduce the precision of mathematics into chemistry. I have been able to show, largely through his unpublished manuscripts, specifically what he meant when he claimed to solve problems in chemistry by reasoning like that of mathematicians. In some of his later manuscripts and published papers, he represented the relations between the substances that entered a reaction and its products in the form of an algebraic equation.[170] One of the consequences of this practice was that in discussing fermentation in his *Traité*

168. Ibid., 201–204.

169. Henry Guerlac, "Chemistry as a branch of physics: Laplace's collaboration with Lavoisier," *Historical studies in the physical sciences, 7* (1976), 193–276.

170. F.L. Holmes, *Lavoisier and the chemistry of life: An exploration of scientific creativity* (Madison, 1985), 280–283, 289, 341–345, 371–372, 394, 396.

171. Lavoisier (ref. 29), *1*, 141.

élémentaire he wrote:[171]

> One must suppose in every [experiment in chemistry] a true equality or equation between the principles of a body that one examines and that one can obtain through analysis. Thus since must of grapes [grape sugar] gives carbonic acid gas and alcohol, I can say that *must of grape = carbonic acid + alcohol.*

This passage is regarded as the prototype of the chemical equation.[172] In Lavoisier's hands, however, the chemical equation conveyed no more information than was contained in the balance sheet form, which historians trace less to mathematics than to commercial bookkeeping practices. Thus the model of mathematics does not appear to have changed Lavoisier's practice of chemistry deeply, but did leave a permanent mark on the symbolic representation of the results obtained through his mode of chemical experimentation. Historians who pursue the view that Lavoisier sought to model chemistry after experimental physics must similarly identify the specific types of physics experiments that he may have had in mind and show how he imitated these experiments in the design of his chemical experiments. Otherwise the claims that Lavoisier modelled chemistry after physics or that he was a physicist rather than a chemist will remain in the realm of rhetoric. They will not constitute a serious challenge to the persuasive argument given by Maurice Dumas thirty years ago, that in his approach to the problem of heat as well as to other basic questions of composition, Lavoisier "acted like an eighteenth century chemist."[173]

Neither of the remaining two global historical alternatives that Donovan justaposes—that Lavoisier reconstructed the entire science of chemistry from the inside, or that he elevated what had formerly been only a practical art to the level of a science—is compatible with the point of view that I have developed in the preceding lectures. Donovan is certainly correct to say that Lavoisier and his supporters sometimes wrote as though they believed they were creating a wholly new science, that they were setting new methodological standards, or that the chemistry of the past was so ridden with

172. Partington (ref. 64), 3, 480.

173. Maurice Daumas, *Lavoisier: Théoricien et expérimenteur* (Paris, 1955), 157–165.

error that a fresh start was required. These were subjective views natural to the leaders of a reform movement, especially when it encountered opposition. Lavoisier's perspective on the relation between his work and the prior state of chemistry varied with time and circumstance, and cannot be encapsulated in any of the numerous statements he made on the subject. When pressed by what he took to be obstinate defenders of views he believed he had supplanted, he could be harsher in his overall judgment of the past than when he wrote in other moods. If my portrayal of earlier eighteenth century chemistry is valid, then the chemical revolution cannot have overturned the science of chemistry as a whole, or have established a science for the first time. It must instead have transformed certain extensive areas of a science whose scope exceeded those areas. Lavoisier himself recognized, within contexts conducive to such recognition, that what he had transformed represented only large parts within a larger whole.

One such context was the one in which Lavoisier himself framed the presentation of his general theory of the nature of acids in 1777. He began his paper by noting that the old chemists, having reduced bodies to oil, salt, earth, and water, believed they had reached the limits of analysis. "In the measure that the art made new progress," however,[174]

> the chemists who succeeded them learned that the substances that they regarded as principles were susceptible of further decomposition; and they recognized successively that all of the neutral salts, for example, were formed by the union of two substances: any acid, and a saline, earthy, or metallic base.
>
> From that came all of the theory of neutral salts, which has fixed the attention of chemists for more than a century, and which is so perfected today that one can regard it as the most certain and the most complete part of chemistry.
>
> In accordance with this state in which the science of chemistry has been transmitted to us, it remains to us to do for the principles comprising neutral salts what the chemists who were our predecessors did for the neutral salts themselves: to attack the acids and bases, and to push back by another degree the limits of this type of chemical analysis.

174. Lavoisier, "Considérations générales sur la nature des acides et sur les principes dont ils sont composés," *MAS*, 1777 [publ. 1780], 535–547, on 535.

In this passage Lavoisier not only viewed himself as continuing along an investigative pathway that his predecessors had followed for a century or more; he implied as well that there were various "parts" to the science of chemistry, at least one of which was already mature. He aimed to do something similar to what they had done for a part of chemistry that had not yet received such treatment.

This passage reflects Lavoisier's attitude at a point when he was on the verge of his drive to replace the theoretical structure of phlogiston chemistry with one based on his new general theory of combustion. Perhaps he had not yet broadened his attack sufficiently to envisage himself reconstructing the whole science of chemistry. The organization of his *Traité élémentaire de chimie* reveals, however, that in 1789, when he was ready to codify the results of a completed revolution, he retained a similar view of the relation between what he believed he had created and what had already existed when he entered the field.

Part one of the *Traité* is entitled "On the formation of aeriform fluids and their decomposition; on the combustion of simple bodies and the formation of acids."[175] When he wrote in his preliminary discourse to the *Traité* that he had found himself "forced to follow an order which differs essentially from that which has been adopted up to the present in all of the works of chemistry;" when he went on to his famous precepts about proceeding "from the known to the unknown;" and when he discussed the disorder, confusion, and false judgments introduced into the science by the lack of a logical language, he applied these remarks only to this part of his work. "It is this part alone," he wrote, "which contains the totality of the doctrine that I have adopted."[176] Thus the new doctrine that he had adopted and that he sought to present in his *Traité* in an "elementary form" was not a whole new system of chemistry, but of aeriform fluids, combustion, and the formation of acids.

The second part of the *Traité* is entitled "On the combination of acids with salifiable bases, and on the formation of neutral salts."[177] Of this part he wrote in his preliminary discourse, "This second

175. Lavoisier (ref. 29), *1*, 1.
176. Ibid., *1*, vii–xxii.
177. Ibid., *1*, 189.

part contains nothing for which I can claim credit [*qui me soit propre*]; it presents merely a very concise summary of results extracted from various works."[178] As he had in his article on acids a decade earlier, therefore, Lavoisier treated the subject of neutral salts as a part of chemistry in which he could add little to the achievements of his predecessors.

There are no affinity tables in the *Traité élémentaire*, and Lavoisier did not discuss affinities in the body of the work. In his preliminary discourse he explained why he would not treat

> the part of chemistry the most susceptible, perhaps, to become one day an exact science; it is that which treats chemical affinities or elective attractions. M. Geoffroy, M. Gellert, M. Bergman, M. Scheele, M. de Morveau, M. Kirwan and many others have already assembled a multitude of particular facts...but the principle givens are lacking, or at least...not yet precise or certain enough...to become the fundamental base for...so important a part of chemistry.

Moreover, he acknowledged, his own self-esteem might be involved. "M. Morveau is at this moment ready to publish the article on affinity in the *Encyclopédie méthodique* and I have ample motive to fear working in competition with him."[179]

This disclaimer seems to have misled some historians. Melhado, for example, writes that affinity theory was "of little interest" to Lavoisier "throughout his career."[180] Daumas, however, gave compelling evidence in 1955 that Lavoisier was considerably interested in the subject. In his more specialized memoirs, Lavoisier invoked affinities in appropriate contexts, and he himself constructed a table of the affinities of oxygen for twenty-five different substances. He was accurate in his assessment of the difficulties surrounding affinity theory, and optimistic in his expectation that his colleagues might resolve them in the future. It was reasonable to exclude a

178. Ibid., *1*, xxix. It is not literally correct that Part 2 of the *Traité* contains nothing for which Lavoisier could claim credit. That is because he did not "follow strictly the plan" that he had formed for the parts of the work, but inserted the tables of simple substances, one of the central features of his "revolution," into Part 2 (ibid., 2: 189–190). That modification does not, however, obviate the general point made above.

179. Ibid., *1*, xiii–xiv.

180. Melhado (ref. 167), 200.

subject in such a state from an elementary textbook. Daumas has shown that for the more comprehensive *Cours de chimie expérimental* that he planned to write in 1792, Lavoisier intended to include a discussion of affinities. According to Daumas, Lavoisier had a critical appreciation of the faults inherent in existing affinity tables, but approved the general direction of contemporary research and had no significantly original ideas on the subject.[181] In his comments on affinities in the *Traité*, therefore, as in his acknowledgment that Part 2 on neutral salts lacked novelty, Lavoisier recognized that his own leadership did not extend to all parts of chemistry.

Historians have long sought criteria by which to differentiate Lavoisier's chemistry from earlier chemistry by something deeper than the substitution of one theory of combustion for another, or any other of his views on specific questions of chemical composition. Donovan identifies this distinction with Lavoisier's use of experiment as the means to acquire *theoretical* knowledge. That this aim was a departure implies that earlier chemists did not have similar aims, though Donovan concedes that earlier chemists such as Hermann Boerhaave and Joseph Macquer were well aware of the importance of both theory and practice.[182] I agree wtih Donovan's view that we must distnguish between the programmatic recognition of the importance of theories and the successful construction of theories through investigative pursuits. The case of neutral salts therefore is critical. Noting that Lavoisier acknowledged that the study of neutral salts was well developed and could be "incorporated into his reorganization of the subject as a whole," Donovan refers to this area as a system "of classification and explanation."[183] Lavoisier himself referred to the "*theory* of neutral salts," however, and I believe he was justified in granting it the status of theoretical knowledge.

No single chemist worked out and proposed the theory of neutral salts in the systematic way that Lavoisier introduced new theories of combustion, acidity, or caloric. It emerged gradually out of the background of seventeenth-century experiments and concepts.

181. Daumas (ref. 173), 171–177.
182. Donovan (ref. 165), 9.
183. Ibid., 39.

When Homberg outlined the classification I described in my second lecture, he was consolidating a view that chemists already tacitly employed, rather than proposing something novel. It required the contributions of successive chemists over the next few decades to develop a comprehensive, well-structured theory of neutral salts thoroughly grounded in experimentation. If we cannot identify this theory strongly with a single chemist, there is nevertheless no doubt that it was by Lavoisier's time far more than a localized set of explanations. It was a broadly coherent theory around which an investigative discipline was structured.

All of this does not deny that Lavoisier was the most probing, the most systematic, and the most critical theoretician who had up until his time entered the field of chemistry. Nor does it deny that his qualities as a theoretician benefitted significantly from his contacts with experimetal physics and with mathematics. He did impart to chemistry a more inclusive, more coherent, and more durable theoretical structure than it had ever previously possessed. He was able to do so not because his mode of reasoning, or his program of research differed in transcendent respects from that of his predecessors and contemporaries in chemistry. He surpassed their achievements because he fixed his attention on certain critical problems pertinent to both physics and chemistry, because he organized his busy life so as to pursue these problems steadfastly but flexibly, and because he brought to these problems the qualities of an exceptionally creative and an unusually critical mind.

2. SALIENT FEATURES OF THE REVOLUTION

Between the narrowly conventional interpretation of the chemical revolution as the overthrow of the phlogiston theory and the most expansive efforts to embrace within it the construction or reconstruction of an entire science, or even the advent of a profoundly new way to do science, there has lain a large arena for historical debate about the scope, the temporal boundaries, and the conceptual or methodological features that define the revolution.[184] In

184. For a succint summary of these questions, see Maurice Crosland, "Chemistry and the chemical revolution," in G.S. Rousseau and Roy Porter, eds., *The ferment of knowledge* (London, 1980), 389–416.

1890 Marcellin Berthelot, the first historian to apply the phrase "La révolution chimique" to Lavoisier's achievement, summarized the event as follows:[185]

> In that epoch science was, in fact, transformed by a considerable revolution in the ideas reigning up until then, I would say not in chemistry alone, but in the totality of the physical and natural sciences. The constitution of matter was established according to new conceptions: the old doctrine of the four elements, which had remained since the time of the Greek philosophers, fell. The composition of two of them regarded as simple, air and water, was demonstrated by analysis; earth, a unique and confused element, was replaced by the multitude of our empirical simple bodies, defined with precision. Fire itself changed its character; it ceased to be viewed as a particular substance, to become a pure phenomenon.

From "these new premisses" flowed further consequences for the composition of living creatures and their relation to the atmosphere; for medicine, hygiene and industry; and for philosophy and metaphysics. Berthelot saw the event as a revolution paralleling the contemporary social revolution in France not only because of the nature of these changes, but also because of their rapidity. "They had not been effected gradually, by a slow unfolding of the years and the accumulated work of several generations of thinkers and experimenters. No! They were, on the contrary, produced suddenly: fifteen years sufficed to achieve them." Moreover these changes were, in Berthelot's opinion, not due to collective work but to the persevering efforts of one man.[186]

According to Berthelot, Lavoisier's revolution rested on "a fact, a theory, and a practice." The "fact" was the discovery of the gases by his contemporaries. The idea was a new notion of heat introduced at the same time by physicists. The practice, initiated by Lavoisier himself, was "to weigh exactly all of the products of chemical reactions; not only to weigh the solid or liquid products, as one had always done, but especially to weigh the gaseous products."[187] Despite these methodological and empirical foundations,

185. M. Berthelot, *La révolution chimique: Lavoisier* (Paris, 1890), 1–2.
186. Ibid., 2–3.
187. Ibid., 41.

the revolution was for Berthelot primarily a transformation in ideas.

Even from this condensed summary, we can see that Berthelot did not fix on a single defining attribute of Lavoisier's achievement, nor did he claim that it involved a total reconstruction of a science. The chemical revolution pervasively affected all of chemistry, the physical and natural sciences, but not through some deep underlying or unitary transforming principle. It dealt with problems that were so central, and the ramifications of its basic conclusions were so manifold, that "They touch on questions that interest civilization and human life in the highest degree."[188] Some aspects of Berthelot's definition of the chemical revolution require qualification in the light of later historical work, and the heroic tones in which he portrayed Lavoisier are at odds with our more skeptical age. Nevertheless his overall characterization of the epoch-making event includes most of the elements that subsequent historians have explored in further detail.

This is not the place to review the general development of historical interpretations of the conceptual structure of the chemical revolution since Berthelot. I shall mention only two examples of significant reassessments. In 1968 Robert Siegfried and Betty Jo Dobbs published an article on "Composition: A neglected aspect of the chemical revolution," in which they argued that the prevailing emphasis on the overthrow of the phlogiston theory and the introduction of weight measurements as the hallmark of the revolution had obscured the central importance of the inversions that Lavoisier's new system of chemistry caused for previous views of composition. Substances such as air and water previously assumed to be simple were now viewed as compound, whereas others such as metals that had appeared as compounds of phlogiston with something else were now regarded as simple. For the small traditional number of metaphysically fundamental principles, Lavoisier substituted an indefinite number of operationally identified elements whose status was always subject to future revision. Siegfried and Dobbs argued that this change provoked a crisis in chemistry that was not resolved until the advent of Dalton's atomic theory. They view the chemical revolution therefore not as an epoch culminating in 1789 with Lavoisier's *Traité élémentaire,* but as one beginning the

188. Ibid., 44.

appearance of that treatise and ending only two decades later with the appearance of Dalton's *New system of chemical philosophy*.[189] From the passage in Berthelot's book previously quoted, we can see that there were historical precedents for a compositional interpretation that Siegfried and Dobbs themselves overlooked. Nevertheless, they pursued the implications of this view sufficiently to pose a radical challenge to the customary definition of the boundaries of the chemical revolution.

Several scholars, including particularly J.B. Gough, have elucidated the fundamental and germinal role of Lavoisier's theory of the gaseous state. The view that all matter can exist in the solid, liquid, or gaseous state was a fundamental departure from the prevailing eighteenth century belief that air was the sole "elastic fluid." Gough has shown that the new conception was not a consequence of Lavoisier's work on combustion, as often supposed, but one he had adopted in outline even before he took up the problem of combustion. Gough has traced the background for this view to the writings of Johann Theodor Eller and A.R.J. Turgot.[190]

These and other recent scholarly efforts have enriched our understanding of the nature of the chemical revolution. Their authors tend, however, to interpret their findings less as expansions of our sense of the complexity of the event than as new candidates for the central defining thread of the revolution. Thus even while acknowledging that their new perspective complements instead of replacing the older views, Siegfried and Dobbs asserted that "these compositional changes are of sufficient importance to require that the 'chemical revolution' be redefined in terms of them, rather than with the overthrow of phlogiston and the triumph of the oxygen theory."[191] In contrast, Gough maintained that "the problem central to the beginning as well as the whole of the subsequent revolution in chemistry was not combustion, but the nature of vapors and the air and their relation to one another—in other words the problem of the gaseous state."[192] Lavoisier himself would not, I believe, have

189. Robert Siegfried and Betty Jo Dobbs, "Composition: A neglected aspect of the chemical revolution," *Annals of science*, 24 (1968), 275–293.

190. J.B. Gough, "The origins of Lavoisier's theory of the gasesous state," in Harry Woolf, ed., *The analytic spirit: Essays in the history of science in honor of Henry Guerlac* (Ithaca, 1981), 15–39.

191. Siegfried and Dobbs (ref. 189), 292.

192. Gough (ref. 190), 15.

recognized such distinctions of priority among the various problems that he pursued. At various times one or another of them became more central to his immediate concerns; but he treated each of them as an essential component in a network of interrelated problems that constituted his investigative enterprise.

Concurrent with re-evaluations of the overall character of the chemical revolution, there have been since Berthelot's history numerous scholarly efforts to trace more fully Lavoisier's course toward its realization. As in the historical treatment of the background preceding Lavoisier, there has long existed a canonical story of the stages in his progress. The "first phase" begins with his initial experiments on the calcination of metals and combustion of phosphorus and sulfur in 1772 and moves successively to the discovery of oxygen in 1775, the oxygen theory of acidity in 1776, and the composition of the atmosphere and general theory of combustion in 1777. The middle phase, lasting from 1782 to 1785, features the first calorimetric measurements of the heat released in chemical processes, the caloric theory of heat, and the discovery of the decomposition of water. The final stage leads from Lavoisier's general attack on the phlogiston theory and the first conversions of important French chemists to his views to the formulation of a new nomenclature in 1789 in Lavoisier's *Traité élémentaire de chimie.* Further strands of Lavoisier's work, such as that on the composition of plant matter, fermentation, and the theory of respiration, are most often treated less as integral components of the chemical revolution itself than as applications of its consequences to the fields of organic chemistry and physiology. These topics recur, with some variations in emphasis, in works from Berthelot's *La révolution chimique* to the brief but authoritative biography published by Guerlac in 1975.[193] The most conspicuous recent modification of the earlier framework has been the insertion of Lavoisier's germinal views on heat and the gaseous state as a prelude to the former "first phase."

The aim of much of the detailed scholarship on Lavoisier over the past century has been to refine the interpretation of these

193. Berthelot (ref. 185), 45–188; Henry Guerlac, *Antoine-Laurent Lavoisier: Chemist and revolutionary* (New York, 1975). The content of this book appeared originally in the *Dictionary of scientific biography, 8,* 66–91.

accepted nodal stages in his scientific career. Historians have focused most intently on the origins of the "first phase:" that is, on the genesis of his experiments on combustion and calcination. This topic continues to attract much attention, not only because of the historical fascination of germinal events, but because the sparse surviving record of Lavoisier's activity during this period stimulates the search for strategic previously unknown documents and challenges the historian to build a coherent interpretation upon clues that are suggestive but inconclusive. The significant earlier contributions of A.N. Meldrum, Max Speter, and others to this question were largely superceded in 1961 when Guerlac's book *Lavoisier: The crucial year* provided a masterly synthesis of the evidence then available to account for how Lavoisier became interested in and carried out his first experiments on combustion.[194] As plausible and well-argued as Guerlac's reconstruction was, however, it left room for doubts on certain key points and gave rise to a number of alternative interpretations.[195] Most recently, Carl Perrin has found several more critical unpublished manuscripts that he dates to this period. He has used them to construct a narrative of the first months of Lavoisier's work on combustion that is richer and more persuasive than any of the previous accounts.[196]

Recent historical studies cluster also around each of the problems associated with the subsequent landmark steps along Lavoisier's scientific journey.[197] What remains unwritten is a comprehensive narrative connecting these episodes into the continuous investigative pathway that Lavoisier pursued during the two decades between 1772 and his death. There are abundant documentary sources to support such a study. In addition to the numerous major and minor scientific papers he published, his laboratory notebooks from early 1773 until 1788 are intact.[198] Although regarded as

194. Guerlac, *Lavoisier: The crucial year* (Ithaca, 1961).

195. Robert E. Kohler, "The origin of Lavoisier's first experiments on combustion," *Isis, 63* (1972), 349–355; Robert J. Morris, "Lavoisier on air and fire: The memoir of 1772," *Isis, 60* (1969), 374–377.

196. C.E. Perrin, "Lavoisier's thoughts on calcination and combustion, 1772–1773," *Isis, 77* (1986), 647–666; "Continuity and divergence of research traditions: Lavoisier, Stahl and Hales," *Osiris, 4* (1988), 53–81; and "Document, text and myth: Lavoisier's crucial year revisited," *British journal for the history of science, 22* (1989), 3–25.

197. See the comprehensive topical bibliography in Guerlac (ref. 193), 143–163.

198. Lavoisier, Cahiers de laboratoire, Archives de l'Académie des Sciences, Paris.

incomplete, I believe that they represent the primary record of his experiments throughout that period. They have lain strangely unexploited by Lavoisier scholars, who have sometimes relied on an abstract of their contents provided by Berthelot that is an inadequate substitute for the original.[199] In addition, Lavoisier left many informal memoranda recording ideas in their early stages of development, and successive drafts of most of his scientific papers that often reveal further stages in the evolution of his views. From these materials it is possible to construct a dense, illuminating account of his investigative trail, showing the interplay between the conceptual frame that grew from a rudimentary cluster of ideas at the beginning to a tightly structured system at the end, and the intense experimental activity that both tested and gave rise to successive stages in the development of his ideas. I have utilized this approach to follow several major strands within the broader stream of his multifaceted investigative enterprise—those related to biological problems.[200] Such an approach can be exended to encompass the entire network of his scientific activities.

A narrative of Lavoisier's investigative pathway that is both broad in scope and finely detailed would, I am certain, significantly modify the familiar images of his achievements and his scientific personality. It would embed the evolution of his ideas, to which so much historical attention has been given, into a context of ongoing experimentation. It would show that the various facets of his investigation were too intimately interconnected to fit the neatly bounded steps or phases into which the discussion of his work is customarily divided. It would soften the edges of his presumed methodical style and would reveal an experimentalist less systematic but more opportunistic and more spontaneous than the Lavoisier known through his polished public performances. It would portray a scientist who had remarkable insight and foresight, but possessed less prophetic knowledge of the goals he would eventually reach than he is sometimes portrayed as having. It would, I believe, reveal a more human, more vulnerable, but in my opinion no less impressive historical figure than the one whose achievements are

199. Berthelot (ref. 185), 225–310.
200. Holmes (ref. 170).

customarily discussed from more distant perspectives.

Historical interpretation is partially shaped by the format within which it is communicated. Older views of Lavoisier and the chemical revolution were formulated largely in books such as that of Berthelot and the biography by Edouard Grimaux, written in 1888,[201] which served as a model for subsequent biographies and remains a major source for information. Most of the recent reassessments of the subjcet—with the conspicuous exception of *Lavoisier: The crucial year*—have appeared in the form of scholarly articles.[202] Such articles fall into two characteristic genres: broad overviews that can be suggestive but are rarely definitive, and sharply focused analyses of crucial nodal problems within a larger picture. The restricted scope of the article form favors the development of one or two strong themes. In the case of the chemical revolution, the current predominance of the article form has encouraged the tendency to pick out certain of its features as central while displacing others toward the periphery, and has fostered debates about which events or themes to give priority that may be artificial.

Whatever boundaries one places on the chemical revolution, it was a complex, multidimensional episode. It can be encompassed only by a narrative broad enough to show how the various thematic strands that historians have isolated as critical factors were interwoven. Scholarly essays that highlight special topics within this complex are important, but are only steps toward a story that must someday be told on a grander scale. The present lecture should be seen in that light: merely as a commentary on certain issues that should be embedded eventually within a more comprehensive treatment. Carl Perrin spent many years preparing himself to write such an account. The masterly quality of the articles he has written on special aspects of his subject makes evident

201. Edouard Grimaux, *Lavoisier: 1743–1794* (Paris, 1888).

202. Two recent books contain extended discussions of Lavoisier's ideas. Melhado (ref. 16), 1, 71–83, 124–141, treats Lavoisier's views primarily as a "background to Berzelius's achievements." His discussion is more successful for that purpose than as a reinterpretation of Lavoisier or the chemical revolution itself. Elisabeth Ströker, *Theoriewandel in der Wissenschaftsgeschichte: Chemie im 18 Jahrhundert* (Frankfurt, 1982), gives a detailed but traditional account of the chemical revolution as the victory of the oxygen theory over the phlogiston theory.

how tragic a loss it has been that he did not live long enough to reach his goal.

3. SPECIALIZATION IN EIGHTEENTH CENTURY CHEMISTRY

To close, I return to the position I outlined earlier, that Lavoisier reconstructed not the whole of chemistry, but only a crucial domain within a larger science. This was not due to limitations in his vision or his chemical knowledge. Intellectually, Lavoisier probably mastered the field of chemistry more fully than anyone before or since him has been able to do. He did not, however, and undoubtedly could not master the experimental methods and strategies necessary to make investigative contributions in all of the existing branches of chemistry. The reason should be obvious from my previous lectures. By the last third of the eighteenth century chemistry had grown into a field of sufficient complexity so that a division of labor was becoming essential. Individual chemists who wished to do significant original work increasingly had to specialize. This view may appear contrary to the conventional portrait of the eighteenth century scientist as a generalized natural philosopher dabbling in widely separated areas. In chemistry one could no longer operate that way effectively. Even Joseph Priestley, the quintessential natural philosopher with diverse intellectual interests, specialized in chemistry in the discovery of new "airs." He became expert in the use of pneumatic apparatus, much of which he had himself devised or improved; but his knowledge of the analytic methods currently employed to study the chemistry of salts was rudimentary compared to that of such masters of the art as Marggraf.[203]

Carl Wilhelm Scheele, perhaps the most gifted experimental chemist of the pre-revolutionary era, spent over fifteen years isolating, identifying, and characterizing the family of acids derived from plant and animal matter, which more than doubled the number of known acids. His first success in isolating tartaric acid in 1768 resulted from an ingenious strategy by which he precipitated the

203. This assessment is based on my reading of extended portions of Priestley's *Experiments and observations of different kinds of air,* 3 vols. (Birmingham, 1790).

acid as an insoluble salt of lime and then displaced it from this salt by adding vitriolic acid to form the even less soluble selenite. Afterward, he applied the same method to try to obtain other suspected plant acids, but they proved to be more difficult problems to solve. Persistently attacking them over the years, he deployed an expanding repertoire of variations on his initial strategy that represents the high point in the application of the eighteenth century methods associated with the chemistry of salts. Gaining further experience and flexibility with each effort, he attained a virtuosity in the use of these strategies that easily surpassed the capacity of other chemists in his time.[204]

Scheele's countryman, Torbern Bergman, ranged widely in astronomy, biology, and geology before he was appointed to a chair in chemistry at Uppsala in 1767.[205] For the next seventeen years, however, he focused his scientific energy intently on chemical analysis, channeling his investigations toward an extension of Geoffroy's table of affinities. The table that he published in 1775 contained 50 columns, each of which had 50 places. About half of these spaces were filled in, all on the basis of experiments that Bergman himself had performed. He acknowledged that he had still a long way to go:[206]

> I am certainly aware of the fact that more than 30,000 choice experiments are required for the now proposed and completed outline, if it is to come anywhere near perfection. But in the meantime I have pondered this with myself, that besides the fact that the course of life is short and a state of good health is uncertain, that I ought to publish the observations made by me up to now just as they are....In any case the labor has only been started and if God allows life, strength and the necessary leisure time, I intend to continue it.

Despite Bergman's phrase, this was hardly leisure time activity. It would be hard to find a more poignant testimony to the specialized concentration that a chemist of Bergman's stature could bring to his task, and that the state of chemistry in the late eighteenth century

204. In an unpublished manuscript, I have reconstructed the detailed course of Scheele's investigation from his published papers and correspondence.

205. J.A. Schufle, "Introduction," in Torbern Bergman, *Dissertation on elective attractions* (New York, 1968), xxv.

206. Torbern Bergman, *Dissertation on elective attractions* (New York, 1968), 25.

could demand for leadership in one of its branches.

In his early years, Lavoisier appeared as something of a dilettante, examining diverse scientific questions that lacked a sustained investigative line. The auspicious portion of his career began when he settled upon a particular problem—the processes that fix or release air—and decided to make it the subject for a "long series of experiments."[207] The problem was broad enough to require vision, and he saw immediately its potential to cause a revolution in physics and chemistry; but it was also specialized enough to require its own array of apparatus and methods. Once he had become an expert in the special methods that he devised for himself, Lavoisier could apply them to an ever expanding range of similar problems. The task demanded, however, the full attention of his considerable experimental talents. Historians who treat the chemical revolution mainly in conceptual terms seldom appreciate the degree of sustained experimental effort that Lavoisier had to exert during the next twenty years to turn his initial glimpse of a future revolution into a reality.[208] It was not only a matter of "clarifying" his ideas about the nature of heat, the gaseous state, the constitution of the atmosphere, or the nature of acids.

Having chosen to stake his scientific career on a mode of experimentation that required him to account completely for the weight of the substances entering a reaction in the weight of its products, Lavoisier had to surmount a myriad of obstacles to make his strategy work. There was no global procedure ensuring success in every application. With each experimental problem he encountered he had to make special judgments about what to measure, whether to repeat a flawed experiment or salvage it by estimating the errors incurred, what to accept as a reasonable approximation to the ideal result he sought. Experiments that appear simple in retrospect were difficult to carry out under the conditions of the time, and often failed because of technical limitations in the instruments and apparatus available.

With so much to do in the particular investigative pathway he had set for himself—and with numerous extra-scientific demands on

207. Guerlac (ref. 194), 228–230.

208. Among Lavoisier scholars Maurice Daumas was exceptional in his appreciation of Lavoisier's experimental prowess. See Daumas (ref. 173), 113–131.

his time—it is hardly surprising that Lavoisier did not also pursue other modes of chemical experimentation. It should not be surprising that he did not observe new airs as Priestley did, except for the one that was crucial to his enterprise, but that he profited greatly from Priestley's discoveries. It is understandable that he discovered no new acids as Scheele did, but that Scheele's new acids enabled him greatly to extend the range of his theory of acids.

Historians have made much ado about the extent to which Lavoisier incorporated the results or conclusions of other chemists into his own conceptual framework. One reads repeatedly of Lavoisier's "debts" to others, as though they should somehow be debited from the sum of his own achievements, and he has been accused of insufficiently acknowledging these debts. It is also sometimes inferred from his use of the experimental observations of others that Lavoisier was primarily a theoretician rather than an experimentalist. He was, in fact, the most innovative experimental chemist of his time; but like other leading chemists, he channeled his investigative activity in certain specialized directions, and did not attempt to duplicate the efforts of specialists in other areas.

Historians who treat it as a mark against him that he "borrowed" from others underestimate both Lavoisier and the chemistry that preceded him. The implication that the founder of modern chemistry would have been a greater scientist if he had gathered for himself all the data, or thought out by himself all of the ideas essential to his revolution, is another reflection of a historiography that has denied to pre-revolutionary chemistry the full status of a science. If we view eighteenth century chemistry as a maturing scientific discipline within which individual investigators had to specialize in order to excel, then it becomes self-evident that in order to build a broad new theoretical structure within that field Lavoisier necessarily drew upon the results of many specialized investigations carried out by other able chemists.

EPILOGUE

The eighteenth century investigative thrusts in the chemistry of salts and in plant chemistry described in the second and third lectures took place within laboratories little changed from those of the preceding century. An array of long-familiar equipment and procedures, deployed in new orders and combinations, sufficed to sustain experimental ventures that greatly changed the investigative structure of chemistry. The end of the long period of inertia regarding the physical appointment of the chemical laboratory was, however, already in sight.

The first major novelties in the equipment of eighteenth century laboratories were the various forms of jars inverted over water to collect "airs." Invented by Stephen Hales and introduced into French laboratories by Guillaume-François Rouelle, these devices were at first simple enough not to alter drastically the overall physical arrangement of a laboratory. In the hands of Lavoisier, however, the pneumatic apparatus soon began to transform the character of the chemical workplace. His earliest apparatus can be seen as somewhat makeshift adaptations of the apparatus of Hales or Rouelle. As Lavoisier advanced through an expanding array of increasingly demanding experimental measurements, however, he had a growing need for apparatus expressly designed and constructed for specific problems. The plates at the back of his *Traité élémentaire de chymie* display a complexity and variety of instrumentation quite unknown in chemical laboratories twenty years earlier.[209]

209. Stephen Hales, *Vegetable staticks,* ed. M.A. Hoskin (New York, 1961), plates 15–17, 160–161; Henry Guerlac (ref. 194), 29–31; Antoine Lavoisier, *Essays physical and chemical,* tr. Thomas Henry, 2nd ed. (London, 1970), plates 1–11, following p. 428; Lavoisier (ref. 29), 2, plates iv-xii, following p. 656. This and the preceding paragraph are developed from my reply to a question raised in the discussion following the first lecture.

In the first lecture I compared the stable traditional chemical laboratory as a setting for chemical investigation with Fernand Braudel's treatment of the Mediterranean basin as a stable physical setting for the social and economic order that dominated the Western world during the sixteenth century. According to Braudel, the boundaries of the Mediterranean world were pierced toward 1600 by the opening of trade routes through the North Atlantic and the economic rise of northern Europe. Braudel argues, however, that this broadening of the economic horizons did not eclipse the Mediterranean world, which managed to adapt to the new conditions.[210] In an analogous way, Lavoisier's new generation of instruments pierced the old physical boundaries of the chemical laboratory, but did not supplant it or replace it with a new institution.[211] The familiar furnaces and distillation apparatus, the vessels for solutions, precipitations and filtrations, the reagent bottles and work benches of the old laboratories continued to have their place in the expanding domain of the laboratories of the nineteenth century. Throughout the long duration of its history, chemistry has been an activity firmly situated within, and fully understandable only in terms of, the enduring physical environment of the prototypical institution of the experimental sciences: the chemical laboratory.

Following the third lecture, on plant chemistry, Christoph Meinel asked whether the developments I had described involved conceptual changes, or whether they were merely operational. I replied that in the gradual shift from the traditional distillation procedures to solvent extractions important changes in conceptions about the nature of the constituents of plant matter were embedded. Especially significant was the clarification of the distinction between the "immediate principles" that can be obtained from plant matter unchanged, and the "ultimate principles" or elements into which these immediate principles can be further decomposed.[212] I had not dwelt on these conceptual issues, however, because of my intention

210. Braudel (ref. 36), *1*, 578–580.

211. This continuity is symbolized in the fact that plates I–III in Lavoisier's *Traité* (ref. 29) depict the traditional apparatus of the chemical laboratory. See below Figure 19.

212. For the significance of this distinction for early nineteenth century plant and animal chemistry, see Joseph S. Fruton, *Molecules and life: Historical essays on the interplay of chemistry and biology* (New York, 1972), 87–95.

to present eighteenth century chemistry primarily from the viewpoint of its investigative activity.

During the discussion of the final lecture, Pietro Corsi and Marco Beretta each stressed the central importance of the new nomenclature in the chemical revolution. Corsi asserted that many contemporary observers regarded the new language of the chemistry of Lavoisier and his followers as its *most* essential feature, one that served as a model for other sciences.[213] In my response I noted that the chemical revolution had different meanings for different people—for Lavoisier, for his followers, for their opponents, and for general Enlightenment thinkers.

These and other exchanges that took place during the discussions highlight the multiple perspectives from which any major scientific development can be viewed. I have centered these lectures upon eighteenth century chemistry "as an investigative enterprise" in part because much of the vigorous investigative activity that chemists pursued throughout that century has yet to be incorporated into general historical treatments of the subject. Underlying this choice, however, is a broader historiographical issue.

Systematic investigation of delimited domains of the natural world lies at the core of the activity we call science. It has not, however, found a place at the heart of our historical accounts of the development of science. Within the historiographical tradition dominant between World War II and the early 1960s, the history of science was conceived primarily as a history of ideas. A more recent movement has sought to define science as a social activity, stressing the organization of scientific communities, institutions, career pathways, and the interactions between science and wider social or political interests. To distinguish these two genres of history of science, historians have posed dichotomies such as "internal versus external" history, or "cognitive processes and social processes." These attempts to demarcate a perceived inner from an outer dimension of science leave no space for a history of investigative activities that are both cognitive and operative; that require both the head and the hand; that employ ideas within concrete physical settings; that are simultaneously intellectual and socially organized.

213. For a further discussion of this position, see Pietro Corsi, *Oltre il mito: Lamarck e le scienze naturali del suo tempo* (Bologna, 1983), 51ff.

The phrase "investigative enterprise" applied in these lectures to eighteenth century chemistry suggests an approach to the history of science that can dissolve the polarities so prominent in our field. Throughout the history of Western culture, individuals who called themselves scientists, or whom we recognize retrospectively as scientists, have engaged in a collective investigative enterprise. The fine structure of that enterprise may be traced along the daily research pathway of a single scientist; but the enterprise extends also through a network of cognitive, operational, organizational, social, and cultural strands linking each investigator to groups of individuals active within or beyond the domain in which he or she works. It extends to the laboratory or other physical institutions without which the investigation could not be carried out, and to the sources of financial and moral support without which it could not be sustained. It extends to the means of communication—whether informal or through societies and journals—without which an investigation could not be reported to others. It includes the pedagogical activity without which investigations could not be continued by successive generations.

The historical study of science as an investigative enterprise incorporates ideas, but does not treat them as having a life of their own. It shows how ideas come into and emerge from sustained explorations of aspects of nature that can be confronted in the laboratory, the observatory, the museum, or the field. It is attentive to the ways in which scientists organize themselves, share power and authority, promote their interests, and award or withhold recognition; but it does not forget that scientists do all these things in the first place in order to facilitate the investigative activity around which they orient their lives.[214] The conception of the investigative enterprise can be broad enough to bring scholars coming from the natural sciences, from intellectual history, from philosophy, and from the social sciences together to participate in a collaborative historical enterprise.

214. An exemplary model for this historical approach is J.L. Heilbron, *Electricity in the 17th and 18th centuries: A study of early modern physics* (Berkeley, 1979).

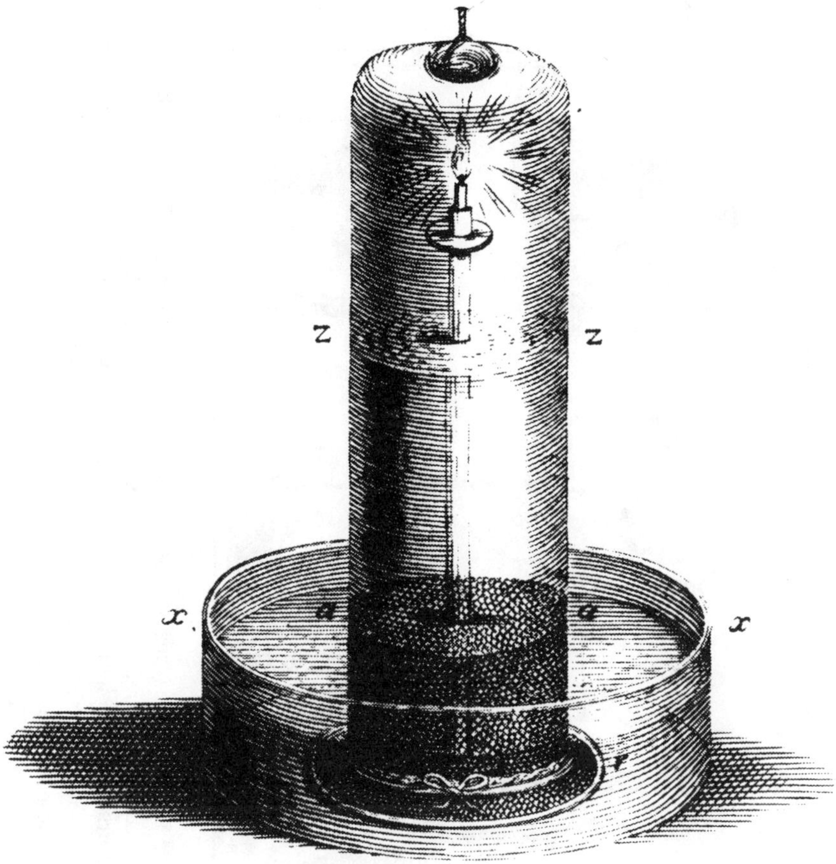

FIG. 15 One of several forms of apparatus used by Stephen Hales to measure changes in volume of air in presence of processes that "fix" or release air. From Stephen Hales, *Vegetable staticks* (London, 1727), facing p. 206.

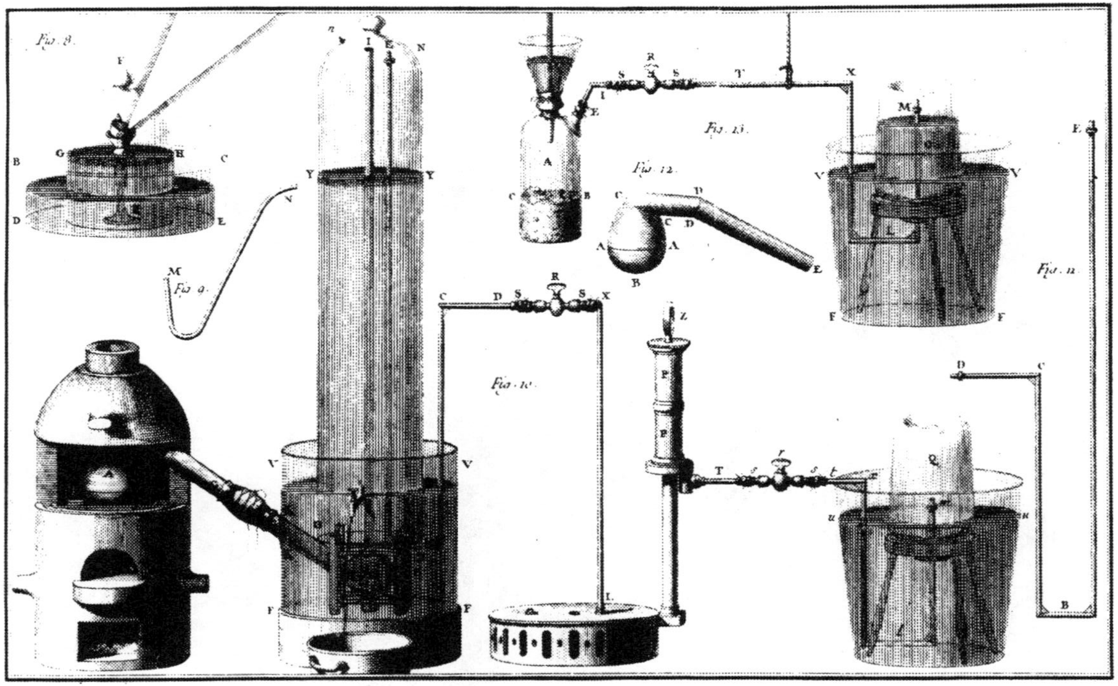

FIG. 16 Apparatus illustrated in Lavoisier, *Opuscules physiques et chymiques* (Paris, 1774), plate II after p. 436. *Fig. 8,* apparatus for reduction of minium using a burning lens. *Fig. 10,* reduction of minium heated in retort (A) constructed of iron; recipient (H) is connected through vertical tube 1 with chamber in which change of volume of air is measured; a vacuum pump (P) is used to raise level of water to Y before beginning the experiment. In these illustrations one can see that Lavoisier's early apparatus was adapted directly from Stephen Hales's pneumatic apparatus and from traditional distillation equipment.

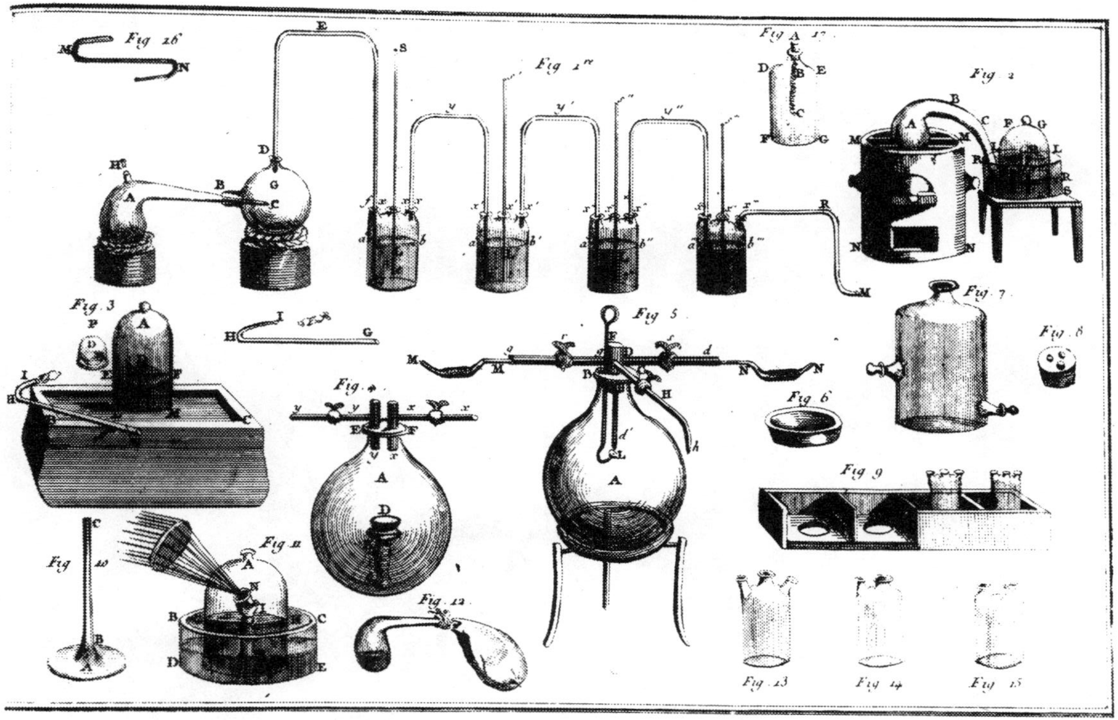

FIG. 17 *Fig.* 3, pneumatic vessel over a marble basin containing mercury. *Fig.* 2, apparatus used in 1776 for analysis and synthesis of air, using mercury. *Fig.* 5, apparatus used for combustion of hydrogen and synthesis of water. From Lavoisier (ref. 29), 2, following p. 656.

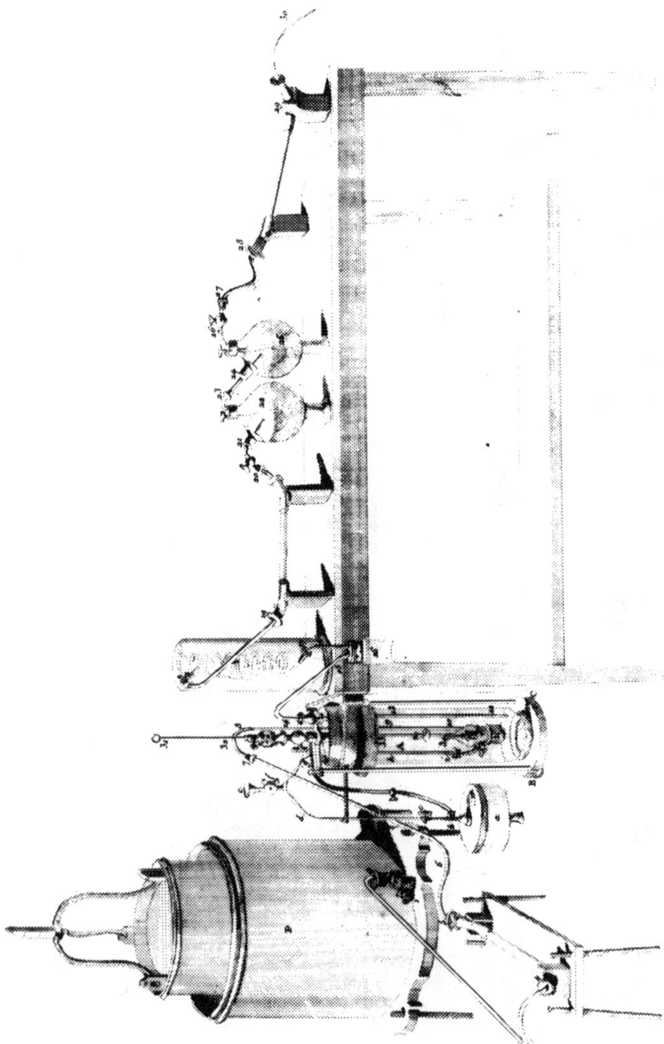

FIG. 18 Apparatus used for combustion of oils. P represents a portion of a gazometer designed to supply measured volumes of a gas. Source as for fig. 17. Comparison of early and late apparatus of Lavoisier shows the growing complexity of his apparatus over the course of his investigative career, and the progressive development of apparatus designed and crafted for specific experimental operations.

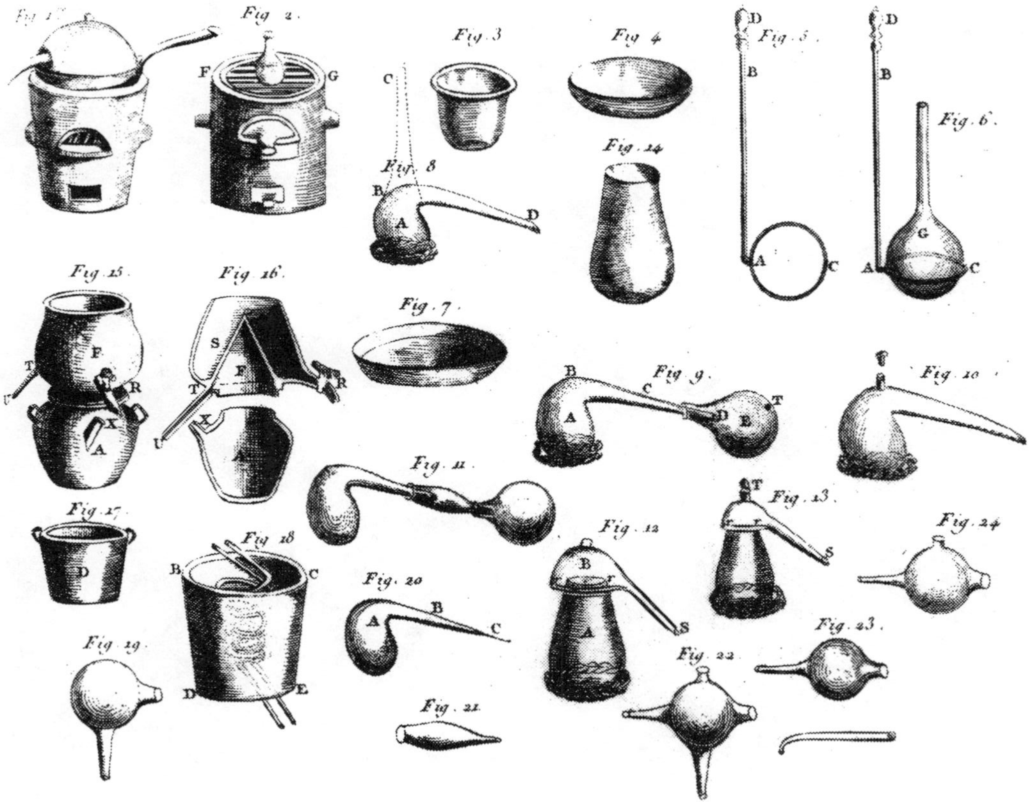

FIG. 19 Traditional chemical apparatus, illustrated in Lavoisier's *Traité*, showing that Lavoisier's new equipment supplemented, but did not replace, that long familiar to chemical laboratories. Source as for fig. 17.

APPENDIX 1

Eighteenth century chemical terms used in the lectures, and their nearest modern equivalents

Absorbent earths: chalk, marble, or clays

Acid of ants: formic acid

Alkaline earths: calcium, magnesium, or barium carbonate

Alum: sodium aluminum sulfate, potassium aluminum sulfate

Argillaceous earth: aluminum carbonate

Calcareous earth: calcium carbonate

Corrosive sublimate: mercuric chloride

Essential salt: a salt characteristic of a particular plant

Fixed alkali: sodium or potassium carbonate

Fixed vegetable alkali: potassium carbonate

Lixivial salt: salt obtained by soaking a solid substance

Magnesia alba: magnesium carbonate

Marine acid: hydrochloric acid

Marine salt (sea salt): sodium chloride

Nitrous acid: nitric acid

Regulus of antimony: metallic antimony

Sal ammoniac: ammonium chloride

Salt of tartar: potassium carbonate

Selenite: calcium sulfate

Spirit of nitre: nitric acid

Spirit of vinegar (acid of vinegar): acetic acid

Spirit of vitriol: sulfuric acid

Spirit of wine: alcohol

Vitriolated tartar: potassium sulfate

Vitriolic acid: sulfuric acid

Vitriols: salts composed of sulfuric acid and a metal

APPENDIX 2

My participation in the International School of the History of Science in Bologna stimulated me to begin studying Italian, with the fortunate consequence that I as able afterward to read Ferdinando Abbri's *Le terre, l'acqua, le arie: La rivoluzione chimica del Settecento* (Bologna: Milan, 1984). It is a very important contribution to our picture of the chemical revolution: one that justifies the contention that the large scope of that event calls for book-length treatments. Abbri's interpretation of Lavoisier's own work, although limited to Lavoisier's published writings, provides fresh insights at many crucial stages. The most significant feature of the book, however, is that it does not restrict its attention to the two chief protagonists in the investigations and the evolution of their positions with discussions of the contributions of numerous major and minor chemists, and physicists—and others without special scientific credentials—to the development of the debate. The central theme of the book is chemical composition. Unlike some recent discussions, however, Abbri does not assert that this is the single dominant thread around which the meaning of the chemical revolution can be woven. He stresses repeatedly the complexity of the development and its multiple overlapping questions.

Abbri's view of eighteenth century chemistry before the advent of the "new chemistry" of Lavoisier is more conventional. He characterizes it as a "Baconian labyrinth" of proliferating theories and empirical investigation (pp. 5–7). The foregoing lectures explain better than Abbri's own account of the eighteenth century background, I believe, how the burgeoning investigative activity of the 1770s and 1780s that he describes could have taken place. That activity presupposes the existence of a well-formed investigative discipline that could not have appeared suddenly, but that was rather the outcome of a long prior development.

INDEX